A TIME AND A TIDE
Charles K. Kao: A Memoir

A TIME AND A TIDE

Charles K. Kao

A Memoir

The Chinese University Press

A Time and A Tide
Charles K. Kao: A Memoir
  By Charles K. Kao

© The Chinese University of Hong Kong 2011

ISBN: 978-962-996-446-7 (Hardcover)
ISBN: 978-962-996-447-4 (Paperback)

THE CHINESE UNIVERSITY PRESS
The Chinese University of Hong Kong
SHA TIN, N.T., HONG KONG
Fax: +852 2603 6692
      +852 2603 7355
E-mail: cup@cuhk.edu.hk
Website: www.chineseupress.com

Printed in Hong Kong

Contents

Foreword

A quarter of a century ago I first put pen to paper, more to write down my memories than because of any notion towards publishing a book. At that time it was more about me and my childhood traumas. I was trying to understand my history. Writing down my thoughts and relating the events that shaped my life helped me to take a step back and analyze matters.

But this informal writing took on a life of its own. Charles was simultaneously writing about his research as well as about how his career was progressing. Somehow our writings became intertwined; we read each other's work and combined the two, which resulted in something completely new. As the project grew we began to think that perhaps, one day, we might publish it.

But the thought of actually assembling a project for publication was unnerving. Neither of us had ever attended a writing course, and I was nervous about receiving criticism of my writing skills. Over the decades I continued to rewrite entire chapters, never quite satisfied with the drafts.

Then one day a publisher approached us to write an autobiography in Chinese. We had neither the time nor the skills to undertake that project, so instead we suggested that they use our English drafts as the basis for an edition translated into Chinese. That autobiography was published some six years ago. The translator did a great job and was not paid at the time— thankfully he accepted royalties over an advance. A couple of years after that a revised edition went on sale in mainland China.

Then the unexpected award of the Nobel Prize for Physics to Charles in 2009 changed everything. Copies of the book both in China and here in Hong Kong began selling like hot cakes! A reprint was ordered and the translator finally received his payment in full.

So much has happened in the years since the Chinese book was first published. In this English version, new chapters have been added, and minor revisions have been made throughout the text for its new English-language audience. Many people, unable to read the book in Chinese, have been pestering me for the English version for a very long time. So, here it is!

Since retiring from academic and public life, Professor Kao has been diagnosed with Alzheimer's, for which there is no known cure. It has remained a largely hidden disease in Hong Kong. People had assumed, in general, that it was simply an affliction of old age. Not much attention was given to Alzheimer's by the medical profession, or the government at large; until, that is, it became headline news. With the publicity and excitement generated in Hong Kong by a Nobel Laureate emerging from one of their own, everyone was finally struck by the reality of Professor Kao's disease. In 2010, with all of this goodwill, we established the "Charles K. Kao Foundation for Alzheimer's Disease," a registered non-profit organization. A website has been set up to inform the public about the aims of the Foundation (www.charleskaofoundation.org). I hope that, with better knowledge and further resources, more can be done to improve facilities for people with Alzheimer's, increase the training of caregivers, and help to relieve the stress for families living with the disease.

A percentage of the sales of each book will be donated to the Foundation, and if you would like to do more, please visit the website.

Gwen Kao
2010

THE NOBEL MEDAL

STUDENTS SIGNING A CONGRATULATORY WALL POSTER

RECEIVING THE JAPAN PRIZE (1996)

DEDICATION CEREMONY OF THE CHARLES KUEN KAO BUILDING (1996)

WITH PREDECESSOR PROF. MA LIN (*middle*) AND SUCCESSOR PROF. ARTHUR
K.C. LI (*left*) AT THE 40TH ANNIVERSARY BANQUET OF UNITED COLLEGE (MAY 1996)

THE KAO FAMILY IN VIRGINIA (1984)

THE TENNIS TEAM (2010)

VISITING WITH GRANDCHILDREN (2009)

PREPARING FOR ANOTHER DIVING ADVENTURE

AT THE POTTER'S WHEEL (2002)

WITH PRINCE PHILIP AFTER RECEIVING THE MEDAL OF THE ROYAL ACADEMY
OF ENGINEERING (1996)

PADOVA HONORARY DEGREE CEREMONY (1996)

ON STAGE IN STOCKHOLM (2009)

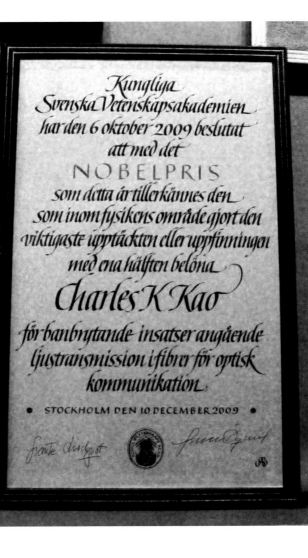

CHARLES AND GWEN KAO ON THEIR WEDDING DAY, 1959

A Wedding

My father was from Kin Shan, a village near Shanghai, and my mother from Pao Shan, some twenty miles north of Kin Shan. They were married in the traditional Chinese manner, after the marriage broker had reported to the respective families that the eight signs under which the bride and groom had been born matched well; in particular, the fact that the families were of compatible social status. Kao Chun Hsian, my father, was the third son of an intellectual well known for his essays and poems. King Tsing Fong, my mother, was the eldest daughter. Both were in their early twenties and well educated for their time.

My father had a scholarly air; an inherited demeanor that allowed him to smile benevolently at the world from behind wire-framed spectacles. And in his world, servants looked after his every pampered need as he grew into a bright future. His bride would most likely be the daughter of family friends; perhaps he would catch more than a few glimpses of her as he grew up—a petite beauty with typical Shanghai features, not particularly curvaceous, with almond eyes, a button nose, and sloping shoulders. Even if they had only met for the first time on their wedding day, I was never able to sense any significant incompatibility. My mother was beautiful and intelligent, while my father, in spite of his liberal training at universities both in China and the United States, was a standard-bearer of Chinese culture and tradition. Over the years they obviously developed a mutual understanding of their respective roles and responsibilities. It was a successful marriage, even if they did begin their married life as strangers. Could it be because the heavenly match of their eight signs assured them success, or had the marriage broker simply been an astute sociologist, adept at maximizing compatibility?

Gwen and I were married in England at an Anglican church in a borough of London. We took the standard vows and signed the marriage certificate. We did not have a marriage broker or a soothsayer to read the eight signs before the families made a mutual decision. Our path seemed a riskier proposition, as we had been friends for two years rather than strangers on our wedding day.

My proposal to Gwen was made several times before she finally answered, "I would really like to have a church wedding." On previous occasions, Gwen simply didn't respond to my proposals. I was becoming accustomed to these periods of deep silence.

Almost immediately after accepting my proposal, a downcast look passed across her face, as she considered the treacherous path that lay ahead. "There are two beautiful churches on the Heath. Let's hope we can find a nice chaplain to marry us," Gwen mused aloud.

Blackheath is in southeast London. Its name derives from the large, sparsely wooded grassland that in ages past might have been scorched black by fires, and around which people settled; over time it grew into a borough of London. Apart from the two churches, the Heath is an oasis of green vegetation in the housing sprawl of suburban London. I was living near the Heath at that time.

The rules for a church wedding required that either the bride or groom be a resident in the diocese, that the couple attend pre-nuptial instruction, and that the wedding bans be read out to the congregation on three Sundays prior to the wedding day. Our first task was to find a place for Gwen to live in the diocese of Blackheath. This turned out to be a relatively easy task. Looking through the classified ads, we came across one listing for a room that was on the right road. The room was in the family home of an elderly couple, and seemed comfortable enough for a one-month stay. It was to be a contingency plan, though we both knew it was likely to be a necessary arrangement for a longer time. The landlords were interested in our plans and graciously lent Gwen their sewing machine, which she used to tailor her wedding dress. My small apartment was a short bus ride away.

The infinitely more difficult task was for me to obtain her mother's permission to marry Gwen. This was the primary hurdle that Gwen preferred not to face. She anticipated that the explosive reaction meted out to her elder sister two years earlier, when the husband-to-be raised the matter with the mother, would be repeated. The tirade was triggered by the fact that, as far as her mother was concerned, the elder son, according to Chinese tradition, must be married before his younger sisters. The eldest sister went ahead with her wedding, but suffered interminable stress and a string of curses, even on her wedding day. The cause of this problem remained, as the elder brother, already well over thirty, was still single, with not a single girlfriend on the horizon. Gwen considered eloping rather than facing the weeks of terrible and vicious verbal abuse that her sister had endured.

"Your no-good boy friend, you think he will take care of you. He is only after one thing—then he will abandon you. He has fed you full of lies. I only think of your welfare. Have I brought you up to be a whore? Have you no respect for your brother? You should have died before bringing such shame to this family!"

Gwen had heard it all two years earlier and knew from experience exactly what would await her. Her brother-in-law had suffered all the indignities, not fully understanding the words yelled at him in an unfamiliar village dialect.

Over the next few weeks, Gwen stealthily moved to her new room, taking a few of her most treasured belongings: some of her book prizes that she had won at school, a pair of favorite shoes, and a few items of clothes. Many years later, her sister told her that everything Gwen had left behind had been immediately thrown into the fire; a glorious conflagration to purge the home of the renegade daughter's existence! Included in this was her stamp collection, a mournful loss of valuable Edwardian stamps that Gwen still searches for in her dreams at night.

On the night she left home for good, I had asked her mother for Gwen's hand in marriage. Her reaction was swift and, as expected, callous. What should have been a celebration was reduced to acrimony and despair. We were prepared for the outcome and, though Gwen had only been able to remove a handful of her possessions, she would not be destitute. The wedding preparations would be our sole responsibility.

The first church we approached was on the edge of Blackheath, perfectly located on a quiet and scenic road. We were disappointed when we found out that Gwen's house was not exactly within the strict boundary of the diocese and St. John's Church. However, the vicar greeted us cordially and suggested that we should go to the other church a short distance away. The second vicar made our visit a short one, indeed. He looked us up and down and said curtly, "We do not marry foreigners." We were left speechless by his open display of racism. Gwen was born in Britain. She is a British subject. He showed us the door, and we trotted back to St. John's where the first vicar immediately responded, "If the other vicar will not marry you, I shall be delighted to do the honor." We were delighted, too, as the church was in a much prettier setting, and England, "this scepter'd isle," showed that it enjoyed pockets of enlightenment.

At that time my cousin Vivian was studying at Avery College, a school of education in London. She was halfway through her teacher training course, and was pleased to act as Gwen's one and only bridesmaid. She proposed sending Gwen's measurements to my mother in Hong Kong in order for a tailor to make Chinese dresses for the two of them. When the dresses arrived Gwen's was too small, as neither the tailor nor my mother could believe that Gwen was any larger than the average extremely petite Hong Kong women. The dresses were duly remade to the correct dimensions, and Gwen and Vivian sparkled at the ceremony.

We found an old school friend—who looked older and more mature—to be the substitute father figure for giving away the bride. Gwen's father had passed away soon after the cessation of hostilities in WWII when she was only twelve years old. Another school friend's brother was my best man.

During this time we both were working fulltime jobs at the same company, which meant that we had to manage our time carefully. We treated the whole exercise like a military operation. We laid out a meticulously detailed timetable for all the things we had to accomplish. We booked the church hall for the wedding reception. We detailed how all the foods were to be prepared and cakes baked in my tiny apartment kitchen. The wedding cake was to be iced and decorated two days before. The sausages, sandwiches, and other snacks were to be prepared the night before. Paper cups and plates, plastic knives and forks, and paper napkins were included on the not-to-forget list. We even prepared the streamers and flowers for the church hall, as well as the packets of confetti for the guests.

We planned our honeymoon so that we would leave from home the morning after. Home was the first floor of a house in Eltham in the south of London. It was the end house of a row of old Victorian brick homes of two floors with high, gabled roofs. The street outside was tree-lined and quiet. The landlord lived in the house next door with his family. His elderly widowed mother lived below us. Some furniture came with the rental: a large, heavy table with four dining chairs in the front room, one well-used armchair, and a small table with two chairs in the back room in which a cooker and a sink were situated. There was no furniture in the bedroom. After finding the place, I moved in just three weeks before our wedding day, sleeping on the floor on a borrowed air mattress. The next weekends were spent buying a bed, bedding, and a refrigerator. I remember that refrigerator well. Still considered a luxury, the small pale green Frigidaire was a vital part of the wedding reception.

Much of the wedding preparation was done there while Gwen commuted from her room each day. On the morning of the wedding, I was to transport the food to the church, while Gwen and Vivian would arrive home early to change. We organized a limousine to transport the wedding party to the church.

I can still recall the rush of adrenaline as dawn broke on the 19th of September 1959. Everything began to fall into place magically. It was a beautiful autumn day—sunny, warm, and calm. The green lawn was speckled with shadows and golden rays angling through the trees as I awaited the arrival of my bride. I was standing outside the church before noon as the guests began to arrive. The organist was playing the requested music. I felt much calmer as I waited for the signal that the wedding party was approaching.

I heard the bell strike twelve o'clock. "Where's the limousine?" I asked my best man, who was looking equally anxious. Had she gotten cold feet?

Later in the day Gwen told me the limousine had gotten so horribly lost she had momentarily thought I'd called off the wedding. To which I responded, "We should trust our resolve. I thought you had chickened out, too." If only mobile phones had existed then, our anxious thirty minutes of doubt would have been eliminated in a second.

As the limousine approached the church and came into sight, my best man ushered me back to my place before the altar. "Did you check whether you have the wedding rings?" I asked nervously. "Of course, I did," as he fumbled through his trouser pockets, before finally finding the rings in an inner pocket of his jacket.

The abrupt change of music signaled that the bridal party was set. I dared not turn around when the wedding march, the one from Verdi's *Aida*, started up. I could imagine Gwen braving the gaze of our friends. I wondered whether she was affected by the absence of her family members. The matriarch had obviously decreed that none of the family should go to the wedding even though we did send the specially printed red invitation cards to all my in-laws: the victimized brother, the married sister, and the younger sister were all absent. Years later, when we confronted them about their lack of courage, they all responded, "Who dared disobey her? It wasn't worth it."

That was the state of the siblings then. Fear of an autocratic mother's wrath proved greater than a sister's need for a single, small gesture of support. I was barely beginning to understand the need for the walls that Gwen had erected around herself. Her inner self was well protected, and it would be my job to begin chipping away at the edifice.

Ironically, many decades later at my son's wedding, his widowed mother-in-law was conspicuously absent from the celebration. She and her daughter had argued violently a few days prior to the event. The mother had stormed out of the room saying she was not going to grace the occasion with her presence. Stubbornly, she held to that decision. Similarly, at the wedding of the daughter of my sister-in-law, the same one who committed the original sin by marrying before the eldest son, the groom's mother did not attend the reception. The estranged father of the groom was in attendance, so his mother would not show her face.

Gwen's eyes were bright and she appeared calm as she stepped beside me, facing the vicar. She smiled like Helen of Troy. Her white brocade dress was stunningly elegant. I had asked earlier whether I could see the dress, to which she replied, "You are not to see the dress until we are at the altar. This is the Western tradition. It will bring bad luck otherwise." I suppose we do mix our beliefs every now and then. I note that bridal couples in Hong Kong happily pose for formal photographs together in their finery days before the wedding.

When we exchanged our vows, we firmly believed that we would never go back on our promises. We were madly in love, and of course didn't realize the enormity of our undertaking. Did we really have any idea what it would take to maintain those promises?

The moment I kissed the bride, I was oblivious to everything except for a feeling of joy and fulfillment. We took the initial step together as we turned to face our well-wishers, who were prepared to bombard us with confetti and paper streamers. The confetti was so deeply embedded in our hair that we were still washing it out in Spain three days later.

The rest of the events swept us along. Photos were taken, champagne flowed, the wedding cake—baked and decorated by Gwen—was cut, the food eaten, all amid the speeches and happy laughter of good friends and colleagues.

The clanging of cans drowned out the roar of our old 1937 Austin 7. The top was down so that we could wave thanks to all our friends who had engineered the cans and the "Just Married" sign scrawled across the rear window.

About the same time we were leaving the luncheon, my parents were hosting a dinner reception in Hong Kong in honor of their first daughter-in-law and eldest son. Their congratulatory telegrams had been read during our wedding lunch in London. Such was the pattern for many weddings held in that era for sons living overseas, as long-distance travel was prohibitively expensive.

The first time I met Gwen, was I truly searching for a person to share my life with? The poet in me might write that I traveled halfway around the world, and fate was surely responsible for our encounter. It actually happened in the austere setting of an engineering firm where I had just begun working. I immediately noticed, among the Caucasian male–dominated work force, an Asian female face. I couldn't decide whether it would be rude for me not to say hello to someone who might be from my homeland or if it would give the wrong signal to my new colleagues if I made a beeline to the only female in the place. I decided not to say anything for a few days.

About thirty years later on a wet and dreary Sunday, Gwen submitted a poem describing our first encounter to a Valentine's competition run by the *South China Morning Post* in Hong Kong.

> Midst oily slicks and cable drums,
> He eyes me as the factory hums.
> And sends to me by sleight of hand
> Secret notes, as I nonchalantly stand . . .
> With colleagues in a canteen line.
> We thought our blushes went unseen;
> Unknowingly grows love pristine . . .
> Soon spied by all, shy Valentine!

As my lunch hour was before hers, we often passed by each other in the canteen or on the stairs going in opposite directions. Gwen had not written any poetry since her school days, so she was astounded to learn that she had won, and I was delighted to escape my onerous academic duties to join my wife for a free holiday at the Bali Hilton.

I joined Standard Telephones and Cables, an engineering company dealing with communication equipment, as a graduate apprentice. It was my first job after college, and I was as green as they come. Even though I tinkered with building radios and had worked on experiments for my engineering studies, the industrial environment was an eye-opener. I was given a desk and assigned to a senior engineer, Peter. On the first day Peter welcomed me warmly and assured me that "You'll get use to us and this place. Take it easy and relax."

I have a tendency to get very nervous and tense in new situations. Peter must have noticed my awkwardness. He said, "Here's something to keep you occupied. Go to the store and get yourself some tools—pliers, cutters, and a soldering iron. The storekeeper will help you." I darted out without asking where the store was, though I must have eventually found it.

My introduction to Gwen took place several days later when I finally gathered up the courage to walk over and speak with her. She was shorter than I, with curly hair and glasses. I noticed her first when my group of fellow apprentices was being shepherded around the buildings on a tour of the facility. "I am Charles Kao, a graduate apprentice. I just started working here a few days ago. How are you?"

She was busy at her bench, fiddling with some apparatus inside a small wooden box. "I am fine. How are you? Are you from Hong Kong?" she responded.

"Yes, I have been in England four years now. I just finished my studies. Are you an engineer as well?"

"Well, I was hired as one. There is only one other female engineer in this coil design group. I suppose the men think that women can only do work that resembles weaving and spinning." She laughed. The ice had been broken.

Throughout my university days, I was a lodger in a suburban pre-Victorian house. The widowed landlady and her son accommodated four lodgers. The lodgers were mainly students and single working people. I remained there from my arrival in England until I started working four years later. In my second year, I happily moved into the larger room and Thomas, a schoolmate of mine from St. Joseph's College in Hong Kong, also joined the household. Thomas was addicted to cars. Even as a poor student he had somehow managed to own a car. It was an old Standard, but it ran well. His car addiction had everything to do with the path of my life, since it was at his suggestion that I invited Gwen to a motor show.

I wasn't sure why he had offered me two extra tickets to the car show at Olympia Hall. Had I ever mentioned to him that I was interested in someone at work? At the car show it was great fun for the three of us to sit behind the steering wheels and imagine ourselves as proud owners. We even waited in line to sit in the Rolls Royce and Bentley. I never forgot that new-leather smell.

As we walked through the enormous hall, Thomas kept disappearing and I was left to escort Gwen. And then whenever we were ready to move on, thinking we had lost him, he would appear out of the blue.

Afterwards we had a quick meal at an Indian restaurant, and then Gwen went home. As soon as she left, Thomas turned to me and said: "I've seen Gwen before at some of the Chinese Student social evenings. She is a very popular girl. You should be aware of this, if you are interested in her."

I muttered, "Who said I'm interested? She's just a colleague."

The matter was dropped.

Soon after the car show, I moved into an apartment. I'd decided that as I was working, it was time for a place of my own. My friendship with Gwen remained largely unchanged. We continued to work together and ran into each other at Chinese Student Association social events. Gwen usually tagged along with her brother and older sister. She was sociable and willing to get involved in organizing committees. We gradually got to know each other better through long conversations at these events.

I was impressed by her lively nature and wide range of interests. Eventually I learned some of her family details, such as how they survived the war and how she struggled through her education. I found out that her mother was a superb cook, and that the student community in London knew her as being a generous hostess. Her mother would provide anyone who dropped by to visit, especially female students, with a culinary treat. Her real motive was to lure an eligible young lady for her only son, William. I was occasionally present at these feasts. Gradually, I became "accustomed to her face," as Prof. Higgins pronounced of Eliza Doolittle. At work, I began to make more trips to Gwen's bench, just to talk. She became increasingly alarmed by my appearances, not wanting our friendship to become laboratory gossip.

One day Thomas was driving us someplace. We were crushed into the car in a mass of bodies. It was a decidedly inappropriate time for me to ask out Gwen. She must have felt extremely embarrassed at my public courtship in the crowded car, but I was oblivious to the amused looks. I continued negotiating with Gwen, who kept making different excuses. Finally, she gave in simply so as to not prolong the embarrassing situation. It was a cold winter evening when we went to see Jerry Lewis and Dean Martin's slapstick comedy. She hated it, but was too polite to say anything at the time. Gwen allowed me to hold her frozen hand to keep it warm on the walk home. Electricity ran through our hands as they touched for the first time. Even now, over 40 years later, we remember this and other moments from that time.

Our love truly "blossomed" when Gwen and I attended a summer camp near Chichester, organized by the Chinese Students Association. When we first started to plan for the trip, it sounded like a great way to spend a well-earned holiday after a year of work, but it quickly evolved into a chance for us to finally be alone together, away from work. I borrowed Thomas's car and headed for the camp. For fear of what others might say, we chose to arrive separately. However, after the first day, everyone at the camp knew us as inseparable. We were not unique. A number of campers met their spouses that summer, including two of my future colleagues from The Chinese University of Hong Kong, and the University of Hong Kong. We have a photo from that time of the three women together, including Gwen.

We traveled back to London and our respective homes via Stonehenge. Gwen was in a pensive mood as we walked amongst the gigantic stone pillars. She was worried that I would not understand her mother's archaic view on marriage. Taking her elder sister's case as an illustration, Gwen thought that the punishment meted out for a second occurrence would be more severe, while I argued that her sister actually got off lightly. The mother's bias for the son made Gwen an early disciple of women's rights. She had seen her egg rations during the war saved for the brother; even her father got second-hand treatment. She was afraid I might simply give up, and, at the time, Gwen was not prepared to risk a showdown with her mother over me.

It was difficult for me after our idyllic camping holiday that Gwen had decided to distance herself. I truly didn't understand the situation. She even suggested that we not see each other for six months. Fate came to my rescue. A bus strike had crippled London. Up until then I visited Gwen every day at work and we would go out on the weekends. What was I going to do now that the buses were not running and we had no way to get to work? I decided to visit her home on foot. This would be a show of my devotion and it might even impress her mother. The walk took hours, including a section through the deserted Blackwall Tunnel under the Thames. When I arrived, her mother welcomed me with an air of astonishment and disbelief. She served me tea and cakes and offered to lend me a bicycle so that I could get back home more easily. I cycled back to Gwen's house the next weekend after the bus strike ended. I hoped that my gallantry would convince Gwen that she could trust me.

My attempt worked. Gwen decided to confide in me. "If William's problem remains, I simply will have to face the music from my mother. But are you really my white knight?" she asked.

I replied, "Of course, but is the situation really that desperate? Your mother did lend me a bicycle. Surely she would at least listen to me. We could even introduce someone to William to make things easier."

"I expected you'd respond that way. My mother would be more likely to lock me up and kick you out of the house."

Gwen eventually persuaded me to be prepared for the worst. I agreed that eloping would be a last resort. We also agreed that if we did have to elope, we would attempt to visit her mother until she came around.

It all turned out exactly as Gwen had anticipated. On that fateful night, I broached the subject with her mother. I understood that I should be very direct, and simply said, "I have come to ask your permission to marry Gwen."

Her face fell instantly, almost before I'd finished my request. Then she started screaming: "How dare you want to take my daughter away. How shameless! Get out of my house this instant, or I'll throw you out. You had better not come back or try to see my daughter again. Never."

I couldn't believe what I was hearing. How could anyone react so violently to a simple marriage proposal? Especially since I thought that she had a soft spot for me? I never understood how a mother could be so concerned about her son's reputation. She had been taught that, according to traditional rules, the eldest son was the heir-apparent and must take priority over all others. This meant that he must marry first and that his wife was expected to produce a son to ensure the survival of the family name. Being the head of the household, he also must look after everyone, including his mother. But what if her son never got married? Were her daughters doomed to remain single?

A moment of silence increased the tension in the room, before Gwen found her courage, and said calmly to her mother, "Obviously, you do not approve. I will be leaving with Charles."

"Go, get out of my sight, both of you," she screamed.

Gwen slammed her house key down on the table and we walked out. This was the last time we would see her mother for several months. Each time we knocked on the door, no one would come to open it, though we could hear the occupants moving around inside. We called through the letterbox flap, but we were to be ostracized by the whole family. Each time we came we left a gift and a note, disappointed at the reception. Then one Sunday we found the door ajar. We took the hint and walked in. The reception was icy, but thawing.

"Who are you? I don't have a daughter like you anymore! She died. Who let you in?"

We were genuinely sorry to have made her so unhappy, and were glad to have the chance to apologize. At our next visit we were informed that William was engaged to someone he had met in Singapore. They intended to return to London soon after their wedding. For a fleeting second we thought about taking credit for his good luck. If we had not forced the issue, William would not have been sent off to Singapore to be introduced to a girl who was willing to marry him.

The morning after our wedding, we set off for our honeymoon in Spain. Tossa de Mar was a small Mediterranean seaside resort, a destination for sun-seeking Europeans. It was my first time on the continent. Gwen, who could speak some French and had visited Paris, acted as my guide. We traveled by overnight train from London to Paris, crossing the English Channel on a ferry. In Paris, we changed trains and traveled south to Barcelona.

When we finally reached our destination, the rustic house seemed quaint and welcoming. Two flights of stairs led from the porch up to our room. I asked Gwen to wait for a moment so that I could take up the suitcases and inspect the room. My real motive was to gauge the feasibility of carrying Gwen up to our room. I was obsessed with the romantic scene of the handsome man carrying the blushing bride into the honeymoon suite. I walked Gwen to the top of the first flight of stairs, and then swept her up, somehow managing to swerve into the room without dropping her.

Forty years later, I threatened to carry Gwen to our hotel room where we were celebrating an anniversary. She said, "You'll break your back if you try to carry me at your age, stupid!"

I responded, "You should never forget the Chinese saying, 'When one is older, one must be even stronger,'" as I struggled to replicate my honeymoon feat.

The ancient castle walls of Tossa de Mar, the warm climate and blue Mediterranean, the colorful costumes, the casual and relaxed atmosphere, and the strong aroma of spices and garlic were all ingredients to intoxicate lovers. The white sand beaches strewn with boulders offered many private coves, and we even contemplated skinny-dipping, but were too prudent to actually attempt it.

When the sun was overhead, the sea breezes stopped. Everyone, except "the mad dogs and Englishmen" who "go out in the midday sun," took a siesta. We found this to be a civilized tradition, which we eagerly embraced. Dinner was very late. Even in our little pension the food was delicious. Maybe it was because we were so hungry by that time of night.

We were very proud that our fellow diners took us for a seasoned married couple and not a pair of honeymooners. When people watched us closely enough, they were amazed by how much we always had to talk about. This is something that continues even to this day, which causes some people to question our status as an old married couple.

Spain became one of our favorite vacation spots. We returned there again and again to savor its unique qualities—the magnificent cathedrals, castles, and mosques. We began our married life in Spain and, as it turned out, we were destined to follow in the steps of the ancient Spanish explorers who had set forth to discover the world. On our return to England, with twenty English pounds as our only worldly savings, we set out on a life that would lead us around the world several times.

STUDYING AT ST. JOSEPH'S COLLEGE IN HONG KONG

School

I was left alone in the courtyard of my first real school. The family servant who was also our rickshaw driver walked me in, turned around, and walked back out. Before leaving he said, "Just wait here. Someone will meet you. See those children over there. They are your classmates."

It was a totally alien scene. Kids were everywhere—running, laughing, shouting, and just standing around. I was fascinated. Then suddenly, everyone started to move; they streamed past me heading in every direction. I soon found myself alone in the courtyard wondering why I was still standing there when everyone else had disappeared. I had heard the sound of a bell a few seconds before but, unaware of its significance, I just stood there as if glued to the ground.

"You must be Charles, our new student. Don't just stand there. We've been looking for you."

This woman who had appeared from somewhere behind me held out her hand and I obediently offered mine. "I am your class teacher," she continued, "Tomorrow, make sure to come to Class 8 as soon as the bell rings."

She led me to a room where boys and girls my age were sitting in neat rows, each seated on a bench attached to a desk. They all looked up as I entered. I felt very uncomfortable as they stared at me. They were obviously curious as to how someone could be so late or lost that the teacher had to go and retrieve me. "There is an empty seat in the third row. Sit there."

That was the first time I had ever attended a school and I was already eight years old. Before then I had been taught at home. My younger brother and I studied together, though he was a couple years younger. An elderly Chinese gentleman taught us classical Chinese, and a Filipino man taught us English.

The tradition of home schooling was an ancient form of education still practiced then in China. My grandfather and father were educated that way. The only things I now remember of that time are passages in the four great books of Confucius and a few excerpts from the five great essays. In past centuries, all of these books and essays were memorized by scholars who were studying for the Imperial examinations. Such examinations had been abolished after the establishment of the first Republic in China. This was an intense form of rote learning. The teacher simply listened to our recitations of the passages without ever explaining the meaning of the words.

Master Confucius said, "To learn and on occasion to repeat what one has learnt, is that not after all a pleasure? "

I now interpret this as "Practicing what you have learned generates satisfaction." Of course, I could choose any number of interpretations since the teacher never explained the analects to us. Or take another Confucian saying, "Is it not a happy event to discover something new by revising what you had learned previously?" In other words, "Search and search again, as a way to discover new facts." I believe Confucius defined the modern term "research" long before the general concept was put to use.

During a debate on the importance of rote learning, I said, "If we do not pack our brains with knowledge, particularly the wisdom of Confucius, or the verse of Milton, we will be unable to appreciate what language can communicate."

The development of language is an evolutionary process. We use words and quotations to express thoughts that induce precise or provocative meanings to a listener. The assemblage of words stimulates our thoughts, particularly if the listener has a vast store of knowledge in his or her head. Effective learning, therefore, must include rote learning. Rote learning is only bad when certain teachers insist that examination questions can only have one right answer.

When our English teacher asked us to read stories from *Tales From Shakespeare* by the Lambs, our English was still rudimentary. However, we found we were able to adequately explain the gist of the story in English to the teacher, after we discovered that we could read and better comprehend the same story in a translated version in Chinese. We were bright enough to string English words together in some way until it was a *précis* of the English text.

Luckily my first school was an experimental school established by a group of scholars who had returned from Europe. They were among the first group of students from Chinese universities who had been sent to France to study. They were so impressed by the French education system that they eventually, after a number of years of planning and much preparation, both in China and France, set up a Chinese-French University in Paris and a string of primary and secondary schools in China. My brother and I entered one of those schools in Shanghai.

The district of Shanghai where I lived was under French rule. Unequal treaties, signed by the imperial government of the Manchu Dynasty, ceded parts of Shanghai to the foreign invaders. One area was the French Concession and another region was the International Concession. Both of these concessions were transferred to Japanese rule when Japan and the United States, on opposing sides, entered WWII on December 8, 1941. The people living in the concessions were lucky. They largely avoided the turmoil of the Japanese invasion of China that had started several years before WWII. In 1937, the Japanese overran Shanghai, but the concessions had remained French and International zones. When Japan entered WWII against the Western Allies, the Japanese took over the concessions. It was a peaceful entry as there was no resistance from the native population.

I remember wartime Shanghai as a peaceful and pleasant place. In 1937 the war did cause significant casualties when a bomb, from a crippled Chinese plane, was accidentally released and landed on the Bund, which was always crowded with people. Many were killed and injured. In 1945 Shanghai, an important port under Japanese control, was a target, even though it was not on the grand invasion route of the U.S.-led allied forces, a route that stretched from Midway Island in the Pacific Ocean to Japan. U.S. planes repeatedly attacked Shanghai. The bombings were aimed at strategic locations, such as the airport. My closest moment with this Great War came when I watched an air battle between U.S. and Japanese fighter planes. The loud rattle of machine gun fire was very distinct as we huddled in a semi-basement room, which was designated by the school as a bomb shelter. From the window I was able to see the two planes circling in the sky. Another time, I was standing on the verandah of our house when a plane flew overhead. The plane bore a U.S. Air Force insignia on its wing. Soon afterwards an anti-aircraft gun shell exploded in the lane in front of our house. We could see the spent bullets and the shrapnel on the nearby wall. I was oblivious to the real danger. Whenever the air raid siren went off, my brother and I would put on thin straw mats as armor around our chests and run excitedly around the house.

The school operated normally during the war. I was not aware that many of our teachers were well-qualified university professors. They were the remaining intellectuals who did not leave Shanghai when the Chinese government retreated to Chongqing, the wartime capital of China. The curriculum included the usual subjects, but we had to learn Japanese in addition to Chinese and French. I still remember the French songs: *Sur le pont d'Avignon*, and *La Marseillaise*. I also remember stanzas from La Fontaine's fable on the importance of hard work, *La Cigale et la Fourmi*.

Half a century later I attended a reunion of some of my fellow students in Shanghai. It was the first time a reunion had been organized. My classmate, another Charles, initiated the arrangements from Hong Kong. He had kept up contacts over the years with a few friends from that era. Mostly through word of mouth, the numbers who came to Shanghai or who wrote letters like voices from the past, totaled well over fifty. At the reunion I recited the first verse of *La Cigale*. One of the teachers present was the same teacher who had taught me French decades ago. He was well over eighty now. For a moment, the ensuing fifty years vanished.

The Japanese lessons at school proved useful, too. I can still repeat a handful of phrases well enough for me to be mistaken for a native speaker of Japanese. We all demonstrated to our Japanese teacher, who came from Japan, our national pride, by refusing to pay attention to him. We would even throw chalk at the blackboard to distract him. He would occasionally get so frustrated that he would throw the eraser back at us.

At the other end of the spectrum, our Chinese calligraphy teacher made learning such a joy that everyone tried their best to develop their calligraphic skills. "Let me explain why I circle some with one red circle and others with two red circles. You see this character with two red circles. It is perfect. Your strokes are strong. And look at the whole word. Doesn't it look like a perfectly balanced picture? That's why I think it deserves two red circles for excellence.

"Take a look at another character with only one red circle. It's pretty good, balanced, and the strokes are strong. But if you look here, you appear to have fallen asleep in the middle of making this particular stroke. The stroke stopped where it should not. It was a small mistake. So I have given you only one red circle. You are making great progress. You will do well if you maintain your concentration, let your hand move freely, and put strength into your strokes."

One classmate of mine excelled in calligraphy. He was taller and stronger. And I was one of the least athletic students in the class. For some reason, this boy decided to pick on me. Without provocation, he daubed black ink on my face. I reached for my pen and daubed back instinctively. A cross-daubing battle ensued. The two ink brushes dueled, like sabers. Soon our faces looked like those of the villains in classical Chinese operas! Our teacher was shocked and hurriedly separated the two gladiators. The boy turned out to be the son of a powerful and notorious gangster in Shanghai. I was lucky that we remained friends after the duel.

After several years at school, I felt settled and had made many friends. Even though I was not selected to play on sports teams, I had my own group of friends. Some were my ping-pong partners, but most of them were my friends from various science groups.

We started chemistry experiments in Primary 6, though the subject was not really part of the normal curriculum. Sunny, who later worked as a professor of electrochemistry at Peking University, was my regular partner. Both of our fathers were lawyers and had little knowledge of science. At that time it was possible to purchase chemicals from educational bookstores. We built laboratories in our respective homes, avidly read popular scientific magazines, and tried out simple experiments.

Sunny said, "You know, we can make different types of gases. Oxygen will make fire burn faster and more fiercely. Hydrogen gas will burn itself. Should we make some?"

I said, "Sure."

In the magazines we discovered that passing an electric current through water can make oxygen. The constituent parts of water are hydrogen and oxygen. This is a simple and satisfying experiment. Using two carbon rods as electrodes, oxygen and hydrogen gases will bubble respectively from each electrode. From the electrode connected to the positive terminal of the battery one of the two gases will be formed and bubble out of the water. From the other electrode connected to the negative terminal of the battery the other gas will be formed and bubble up through the water. These gases then can be collected using a couple of jam jars. To maintain the gases in a pure state, the jam jars must be filled with water and carefully inverted under the water's surface of the container where the electrodes are placed, so that absolutely no air is trapped in the jars. The openings of the jam jars must be moved to a position above the top of the electrodes but still under the surface of the water in the container. The experiment begins with the jam jars in position and the battery connected. The gases bubble up into the jam jars and displace the water in them. After the gases have filled the jars, a lid carefully seals each while the openings are still under water. The proof came when the lid was opened and a lit extra-long match was placed at the mouth of the jar. The hydrogen gas explodes while the match next to the oxygen jar flares up brightly. I will leave it up to the reader to guess whether oxygen comes from the electrode connected to the positive or negative terminal. Even now I find it confusing how we assign the two terminals of a battery status as either positive or negative.

Other experiments for making gases required special glassware. We needed a spirit burner, a glass bottle with a neck where a cork can be inserted, a length of glass tubing, and a flask for holding water. First, the glass tube must be inserted through the cork. We had to buy an instrument that could dig a round hole in the cork through which the glass tube could be inserted with a good fit. We learned that the tube had to be bent such that the other end of the tube, when the cork was inserted into the neck of the bottle, could dip into the water-filled flask. This meant that we had to heat the glass tube and bend it into a V-shape. This was more easily said than done. The long glass tube was too long to start with, and we had no idea how to cut it shorter. An attempt at breaking it left broken glass and jagged ends. Another trip to the shop was necessary for some advice. We came back with a triangular file and the knowledge that we could easily snap the tube cleanly if we first scored a groove on it with the file. This was why we needed a triangular file! Eventually, we successfully produced different gases by heating substances in one bottle that then bubbled through the water into the other flask via the glass tube.

More experiments of greater sophistication were done. The titration experiment for making copper sulphate thrilled me. Titration is the process of obtaining a neutral solution of acid and alkaline substances in liquid forms. The blue copper sulphate crystals formed from the neutralized solution, through the forces of nature, glittered under the sun.

Equally exciting was my first invention. I successfully made exploding mud balls by inserting into the moist mud a filling of wet, red phosphorus and potassium chlorate. The argument is as follows: If an easily inflammable substance, such as red phosphorus, is mixed with potassium chlorate, which I knew to be a strong oxidation agent, then a little added friction should cause the mixture to burn. When the mud is moist and like wet dough, it can be made into a ball. If a mixture of the two chemicals is in water, the mixture will be buffered and not burn. This wet mixture can be inserted and then the muddy shell can be closed to form a round ball. When dried the ball could be picked up gently and thrown against a hard surface. It should explode on impact. The experiment was a great success. We used these mud balls as bombs to frighten cats and dogs, just for the fun of it. Looking back now, we were very lucky that we never hurt anyone or ourselves. Red phosphorus can readily turn to white phosphorus; this latter substance can react with acids to give off the poisonous phosphate gas. Moreover, the mixture of phosphorus and potassium chlorate can spontaneously ignite.

This chemical experimentation came to the notice of my horrified parents when I accidentally burnt my brother's hand with hot acid that had spilled. At that time, he was helping me make silver nitrate, a chemical that could be used to make photographic film. I planned to make the chemical by dissolving a silver ornamental plaque in boiling concentrated nitric acid. (By the way, boiling concentrated nitric acid produces toxic fumes and should never be attempted!) The plaque knocked over the ill-balanced glass light bulb, from which I had removed its filament portion, to use as a container for the acid. This was the first time my parents realized the dangerous potential of my experiments, and marked the end of our childish excursion into unsupervised chemical experimentation. By this stage we had acquired a sufficient amount of cyanide to poison the whole city! It had been easy to buy all sorts of dangerous chemicals. We merely instructed our driver to purchase them and he would obediently execute our orders.

After we stopped playing with chemicals, we discovered that we could make a crystal set, a radio that was sold as a kit. The instructional pamphlet was easy to follow. "Assemble the set by following the diagram, and you will be able to listen to a program from a radio station. You will find all the components necessary to make the radio in this complete set. It includes a solenoid coil, a tunable capacitor, a crystal, and a set of earphones and sufficient lengths of wire. A small thin wooden board with pictures of the components drawn on it is provided, so that you can screw the components in place with the screws provided."

"This is great. Can we buy it?" I asked my parents, who were at the shop with me. "Nothing dangerous here," I said.

"If you can make it work, we will let you buy other components to make more radios," my father responded.

For the next few days, I attempted to make the crystal set work. I encountered many difficulties. The most daunting problem was that I heard absolutely nothing when I put on the earphones. I asked my friend Sunny how he was doing.

Sunny said, "First, you have to understand the principle of how the radio works. I have a book here. You should read it. We need to string a long piece of antenna to catch the radio waves. If we slowly turn the variable capacitor, the particular wave that carries the voice will be picked up and the sound that is on that wave can be heard using the earphones. The crystal recreates the sound. If the radio wave still does not work, take out the crystal, throw it on the ground a few times, and try it again."

The next day I told Sunny, "I got nowhere following your advice. I heard nothing."

Sunny said, "We should keep trying."

Over the next few days, I spent more time looking for the crystal on the floor of my room than I spent trying to hear any sound in the earphones. I was getting frustrated.

But persistence eventually pays dividends. Both Sunny and I gained some understanding of the principles of how the radio works. And it was magic when we tuned the capacitor and heard the sound of music for the first time.

This episode must have left a deep impression on me. During my studies, I realized that the crystal I was playing with was actually a piece of semiconductor, the same material that allowed Shockley and Bardeen in the 1940s to invent the transistor at Bell Labs in the United States. The transistor started the Electronic Age. The tip of a metal wire touching the crystal forms a point contact that acts as a diode. A diode can be used to detect signals carried by a radio wave.

Two years after my school's fiftieth reunion, the largest gathering was held in Shanghai. More than one hundred former students and a number of former teachers came. An amateur cameraman caught me on film as I dabbed away my tears. An aeronautic engineer gave an account of how he helped to build the 747 jumbo jet airliner. I recited the first lines of the Gettysburg address.

A school leaves indelible marks on students and I was lucky to be the beneficiary of an excellent school. My luck continued after I left Shanghai in 1949 with my parents to live in Hong Kong where I entered my second school. Little did I realize that it was to be the beginning of a sequence of great changes in my life. At that time, all I knew was that the civil war had spread towards Nanking, the capital of China. The Nationalist forces had lost a major battle against the Communist forces at a strategically important region centered at the city of Hsu Zhou, north of the Yangtze River. Many analyses of the battle have been written, telling how, after a few skirmishes, the morale of the Nationalist forces crumbled, despite their superior numbers and weapons. This was exacerbated by the lack of coordination between the elite forces under the command of different generals, who were concerned only about preserving the strength of their own group rather than winning the battle. The Nationalist army disintegrated through desertion, voluntary withdrawal, or defection to the Communist forces. Only the river separated the invading forces from the fertile and prosperous region south of the river. Of course, many people were worried about the war and its consequences, my parents included.

So, one cloudy day in late autumn, I found myself on board a ship leaving Shanghai. I clearly remember watching the skyline of the Bund from the ship's deck as the ship left the pier, wondering when I would see that scene again. The gray sky looked ominous. My parents and my brother were with me, all under the strain of having to leave a familiar place for the unknown.

When we arrived in Hong Kong, my uncle and his family greeted us. This uncle, T.M., is the youngest brother of my mother. She had another brother and a sister in Hong Kong, both of whom would emigrate to America in later years. T.M. went to Hong Kong to work for the Bank of China in the late 1930s. In 1948, he was well established as the manager of the China Insurance Company, a subsidiary of the Bank of China. Uncle and Aunt, whom I referred to as Number 2 Uncle and Aunt, had two sons and two daughters. One daughter, Vivian, would serve as our bridesmaid years later. They survived the Japanese occupation of Hong Kong and were well known in the Shanghainese community. The family welcomed us and was responsible for introducing us to St. Joseph's College, a Catholic school run by the Christian Brothers. Several of my cousins were already enrolled there.

It was an altogether different type of school. First of all, the teaching was done totally in English except for Chinese and Chinese literature. Secondly, students all spoke Cantonese, which was completely unintelligible to me. I felt lucky that the school rules were very strict and that all students had to speak English in class. This was a great relief since it meant I could communicate. It turned out to be somewhat of a disadvantage for me in the long run, since had I been forced to learn Cantonese then, I would have been able to speak it more like a native now, instead of being heavily accented, which immediately identifies me as nonnative.

In some ways, St. Joseph's College was much like the school in Shanghai. The teachers were devoted and competent, especially the Jesuit Brothers. Although religious teaching was mandatory, the school encouraged and tolerated a large degree of free thinking and personalized development.

During the first three years of study, in addition to learning much of the normal curricula such as English, Chinese, mathematics, physics, chemistry, history, and geography, we were also taught religious courses in which we had to recite the Catholic Catechism and selected passages from the New Testament. It appeared that I still had the opportunity and ability to commit passages to memory, in addition to learning the art of reasoning. I adapted well enough to graduate as one of the top ten students in Hong Kong, in terms of my School Certificate Examination results. This meant that I was able to enter Class 1, the matriculation class for entry to the only university in Hong Kong, aptly named the University of Hong Kong.

Lasting friendships were made during my four years at St Joseph's. Yuan, a boy from Shanghai, was one of the first classmates who helped me out at school. He was my only classmate who could speak Shanghainese. Perhaps he was also the cause of my not learning Cantonese at that time. He made me feel immediately at ease when I was handicapped by my lack of Cantonese.

Certain classroom events are firmly etched in my mind. One such example was the demonstration of Le Chatelier's Principle in chemistry, that is, all natural compound material structures are formed through minimizing the pressures exerted by the constituent elements around each of the individual elements. The chemistry teacher, whom we nicknamed "Little Man" due to his diminutive stature, asked us, the students, to push him. Naturally, he ended up in a corner of the classroom with us in a formation such that just two or three persons were able to push at him directly, while each of them in turn was being pushed separately by two or three persons. This tightly packed configuration is identifiable as the crystalline structure of geometrical regularity.

Little Man was able to instill in me an indelible picture of the shape of crystals. His ensuing explanation was that the close pack formation is a localized region of minimum energy and is known as a region with lowest entropy. Lowest entropy is the expression used in science to imply maximum order or minimum chaos.

The Class 1 students were school administration helpers, so most of us learned to take on responsibilities by looking after the younger students. I never realized the importance of this at the time, but looking back now, those were the times when young students were being groomed to become responsible adults, a neatly designed part of preparation for the real world.

During my primary school days, I was taught reading, writing, and arithmetic. At the same time I was given exposure to a broad range of knowledge, mostly in exciting ways, from my devoted teachers and from interactions with my schoolmates. I learned the skill of memorizing numerous facts and of experiencing how to use tools. These experiences were committed to memory without being precisely explained in a dogmatic form. Even my university education was centered on the acquisition of tools for digging into knowledge deeply and/or broadly (but with linkages emphasized). In this way I was prepared to think independently and with minimum constraints.

To emphasize this point, my first supervisor, when I worked as a new recruit in an industrial company making communication equipment, said to me, "Why are you designing this amplifier based on the elementary principles you learned in college? Those books were simply there to teach you a principle. It is unnecessary to repeat what you have done already. You are supposed to use what you know to create a new design."

For the very first time, I realized that I was not expected simply to regurgitate my knowledge as if it were needed for answering examination questions. In fact, I discovered later that only a small percentage of the material that I studied at the university was relevant for my work.

I have long been intrigued by the following problem. When two persons are conversing, A says, "Meet me tomorrow." It is likely that B understands the statement. The context within which A makes the statement should be sufficient for B to understand explicitly, with little doubt. Yet the actual words used both have precise and imprecise meanings. Did B perceive the tone to be threatening when A had another meaning behind his words? All languages are designed to be suggestive and therefore imprecise, which is why we must approach education nondogmatically. Returning to my own education, I believe I was fortunate to have followed an almost ideal path.

PLAYING WITH FAMILY IN SHANGHAI (*Charles' younger brother is at the far left with Charles next to him on the slide*)

A Step Towards the Unknown

The tradition in a Chinese clan is to live together as one big family. My grandparents lived within a traditional family compound that contained a number of buildings arranged around a central quadrangle. The main building faced south and was occupied by my grandparents and their unmarried children, with my grandfather as the head of the household. The side wings accommodated his married children, each in their own individual quarters. My father was the third son and was expected to move into one of the wings after his two elder brothers were married. Since marriages were arranged through brokers well known to the family, the correct nuptial order of the brothers presented no undue difficulties. The clan usually lived together and dined together. In his house my grandfather, being the first born of his generation, ruled as the guardian of the family name of the clan. He expected that all the younger relatives of his generation and their descendents would be proud of their lineage and would act accordingly to defend the honor of the clan and to keep the name and traditions of the clan scrupulously clean.

My father only moved briefly into the compound with his bride, and then they left almost immediately for Shanghai. There my mother would return to her parent's house, while my father was sent to the United States to further his legal studies. Before returning to Shanghai two years later, he obtained his Doctorate in Jurisprudence as well as a Master's degree in Law. He was immediately summoned to the Mayor's office for an interview for a possible appointment to a special court that had been established to deal with civil and criminal cases involving non-Chinese persons as plaintiff or defendant. This type of court arrangement was demanded by the international community in Shanghai as a means to protect the international citizens. Even though he was only in his late twenties and had yet to practice law, he was appointed to the court as the Chinese judge, sitting alongside the foreign presiding judges. He was given the job, presumably because of his training as a lawyer and his ability to speak and write English. I suspect that at first he was intended as more of a decoration rather than a learned scholar. When I had the occasion to ask him about those days, he relished telling us about a specific murder case, which appeared on the front page of every newspaper and involved the slaying of a well-known socialite. My father's name also appeared in the paper and he instantly became a prominent public figure. He never discussed actual cases with us, maintaining strict confidentiality even several decades after the cases.

It was a turbulent period, during which the last Emperor of the Qing Dynasty abdicated and China was declared a Republic by the revolutionaries led by Dr. Sun Yat Sen. At that point no single leader had sufficient power to take overall command of China. The warlords with their private armies staked out their territories and became regional rulers. Shanghai was under the control of General Chiang Kai Shek, who was declared the commander-in-chief of the army under Dr. Sun. Shanghai was until then still a treaty port, and it was divided into three parts: the French Concession, the Public Concession (ruled by eight European invading countries), and the Chinese section. The Government of the Republic declared the whole of Shanghai within the sovereignty of the Republic of China. The new government was happy to allow the concessions to retain much of the operational governance of the treaty port. Father bought a house in the concession and began his new life as a two-person nuclear family: a thoroughly modern young family no longer under the watchful eyes of parents.

Their first child soon was born—a daughter. Several years later a brother joined her. My mother was petite and birth could not have been easy for her, which made it even more tragic when both of the children caught measles and died before the eldest reached puberty. The young boy was only four years old. This was a year of epidemic, where the virus spread from house to house and children died in the thousands. After two miscarriages I was conceived and mother was in delicate health thereafter. Due to the circumstances of our birth, my brother and I had a cosseted childhood.

By the time our own daughter was three years old, we actively exposed her to all the germs of childhood. Both of our children caught chicken pox before they began school, so we played host to some of their little friends for a chicken pox sharing party. In those days antibiotics and immunizations for the killer diseases of childhood were commonplace, and as parents we had no fear of them.

Father might have been influenced to some extent by his sojourn in the United States where he had witnessed the two-person nuclear family and liked the concept. Nevertheless, his attitude to his children was part traditional and part modern. Timothy and I had a remote relationship with our parents. We were more influenced by our home tutors, who instructed us in classical Chinese, English, and general subjects. We met our parents in a manner more akin to a daily audience with authority figures—as a courtesy, a duty, and for reporting our daily activities. My brother and I schemed our own adventures as we grew up together.

Under the influence of this half-traditional Chinese and half–laissez faire Western approach, Timothy and I developed a sense of self-reliance and an ability to take independent action. With our tutors we trained our minds to recall the best Chinese literature and our multiplication tables up to 9 x 9. Over the years I have enjoyed recalling all sorts of things, including strings of numbers such as Pi = 3.1415926 . . . up to 35 decimal places. Unexpectedly, I came to realize that the classical essays I recited without any understanding left me with meanings I could choose myself. In other words, the meanings conveyed to me by the essays depended on how I chose to interpret them, rather than a specified meaning explained by the teacher. Confucius said, "The purpose of learning is to know how to apply to good use what we learned." I eventually realized these thoughts were usually broader in meaning than the words conveyed. If we wanted to find out exactly what the author was saying, then we must examine the intent and context behind the writing. Confucius's statement opened up my mind to innovate and attach different forms of understanding throughout my own reasoning. In this way, Confucius helped me become a successful engineer.

On December 8, 1941, Japanese troops entered the Shanghai concession. We did not realize that the invasion was part of the surprise attack launched against the American fleet at Pearl Harbor. The Second World War had begun, and we found ourselves under Japanese occupation. The sudden entry into the city of Japanese troops turned Shanghai into an occupied region. Overnight all personnel from the countries that were at war with Japan began serving their time as POWs. The disappearance of his foreign superiors allowed father to close his legal practice and to maintain a low profile. Instead of attending mahjong parties followed by dinners away, or heading off to the theatres for Peking-style operas and movies, or staying late at the office, we suddenly found our parents at home. Our family life became more intimate, and we actually began eating and spending time together.

The food supply was adequate, and life continued peacefully. Shanghai for us children was relatively normal. When Timothy and I ventured out with our parents, we tried to avoid staring at the Japanese sentry guarding the buildings that had been requisitioned and turned into military headquarters by the occupation forces. We had heard so much about the atrocities being committed by the Japanese soldiers, particularly on suspected spies, that we walked past such buildings as quickly and as unobtrusively as possible. The most notorious building was the one located at the far end of the outer Bund Bridge. It was reputed that no one who entered, unless you were Japanese, ever came back out. We imagined hearing the spirits of the dead escape through the windows.

As the war continued, life became increasingly difficult. Supplies were dwindling to the point that the black market became the norm for getting what had been plentiful commodity products. My parents were so frightened about the intensification of the war in Shanghai that we actively considered escaping to the wartime capital, Chongqing. Apart from discussing the possibility while playing games of Bridge or Go, we never actually moved to escape to Chongqing. With their front lines crumbling, we hoped that the Japanese would be too busy to deal with the further occupation of China.

As my parents were more accessible during this period, we benefited from their presence. I often marveled at Father's ability to answer all of my questions. It was a shocking revelation when I discovered one day that he could not answer one of my questions. His infallibility vanished, and he suddenly became a normal human being. What Father said at the time was helpful. "You will find that we have many things we do not know or understand. No one knows everything." This inspired me to seek and to learn.

My parents relied on the teachers, nannies, and other domestic helpers to look after us in our formative years. I was an avid reader of storybooks, especially the story of the "Three Kingdoms." Whether I was in the bathroom or in between lessons given by the teachers, I snatched time to read the exciting battles and clever maneuvers staged by the brilliant strategist Chu Kuo Liang. He helped Liu Pei, a direct descendent of the imperial household of the Emperor of the Han Dynasty, to strengthen the group of people faithful to the Emperor. They were determined to try their best to resurrect the defunct Han Dynasty. The literary merit of the book ranks high in classical Chinese historical novels. It is a literary masterpiece that has inspired many readers to understand the true meaning of compassion and loyalty, and has remained one of the best-known and most popular Chinese novels from the time it was written over a thousand years ago. I cannot imagine how I could have appreciated such a book when I was only seven years old, but I do remember how difficult it was for me to put the book down when I was summoned for other duties. I must have read those massive two volumes more than ten times. I even remember how to answer a trick question set by fans of this novel: What are the given names of the father of Chow Yue, a famous general of the Wu Kingdom, and the name of the father of Chu Kuo Liang?

The answer is: Chow Kee is the name of Chow Yue's father, and Chu Kuo Wu is the name of Chu Kuo Liang's father. The clue is found in the book in a sentence full of puns and plays on words that informs the reader thus, "Kee begets Yue, Wu begets Liang." The trick is in the play of the double meaning of the words, "Kee" and "Wu." Kee means "already." This renders the phrase "Kee begets Yue" to mean "Since Chow Yue has already been born." Wu means "why." It renders the second part of the phrase as "Why should Chu Kuo Liang also exist." The actual sentence expresses the regret of young Chow Yue. He laments that since he was born first why should a Chu Kuo Liang exist who, being his most worthy opponent, always has to outwit him. I found this type of humor amusing and the intellectual contortion exciting.

One particular childhood event has stuck in my mind. One summer day, I trained the sights of my BB gun on a sparrow perched at the top of a lamppost, instead of at the bull's-eye of a target. I pressed the trigger and the bird fell. I had never imagined that I could actually kill something. I had never imagined that by pulling the trigger my action could result in the death of an innocent bird—that it could, in an instant, become a "murdered" bird. Instead of being exhilarated I was disgusted that I had performed such a dastardly act. I sent my little brother to pick up and dispose of the dead bird. I felt guilty. I never fired that gun again to kill. I often wonder why I still remember this event. A psychiatrist might be able to give an explanation. After more than half a century of living, I realize now that life and death, good and evil, mercy and cruelty, are interchangeable. Perhaps I have learned that nothing is inviolate or unchangeable. I have certainly learned that as we evolve nothing remains static.

I never asked Timothy about the dead bird. If I had asked him, he might genuinely have long ago forgotten about it. He had, as it were, performed a service for me. As the younger brother he was used to running my errands. He later verified this fact when I asked him how it had affected him when I left for Great Britain. He responded without any hesitation, "I realized that my life was now in my own hands. I was liberated." We did not see each other for fifteen years. I visited him in Pasadena, California when he was working at Caltech as a post-doctoral researcher in hydrodynamics. Naturally, we felt obliged to bring up the memory of our childhood time together.

"Do you remember our Sunday afternoon conversations? The topics we chose were mostly about the future. We usually sat around the big table that was placed in the middle of the parlor. We did not have air conditioning in those days, but a gentle, warm breeze blowing through the open doors of the verandah made it comfortable and relaxing. The setting was very conducive to idle speculation. On one occasion I remember your question to me," I said.

"What do you want to be, a lawyer or a scientist?" Timothy had asked.

"I want to be a chemist." I replied without hesitation.

"Why?"

"Well, before our parents found out, you and I spent a lot of time working on 'dangerous' experiments in Shanghai. It was absolutely fascinating. Do you remember the blue copper sulphate crystals the day after the titration experiment? Overnight the liquid became a plateful of beautiful blue crystals. I am studying chemistry in school now, and have a better understanding of the process. When I grow up I want to work hard and become a famous chemist."

"I suppose you should write down and lock this statement in a safe place. When you become a famous chemist one day, you can refer to it and say, 'I told you so.'" had been Timothy's reply.

Timothy laughed, "I remember that day very well. And you have yet to realize your original dream."

We both laughed. For an instant we were our small selves again, our laughter a reflection of our childhood naïveté in the face of our continued learning.

CHARLES (*at far right*) ABOARD THE S.S. CANTON EN ROUTE
TO SCHOOL IN GREAT BRITAIN

Sailing into the Wide Wide World

The euphoria of the restoration of China's sovereignty, including the return of all concessions, quickly turned to despair. Civil war between Chiang Kai Shek's Nationalist Party and the Communists fighting under Mao Tse Tung started anew, even before the Japanese troops had left Chinese soil. During the war against the Japanese, the Eighth Route Army of the Communist Party fought alongside the Government forces. After the Japanese surrender, counting on help from the United States, Chang Kai Shek declared that the Communist ideology had been inspired by a group of insurgents, and he vowed to annihilate these rebels. The Nationalist forces staged an all-out attack in an attempt to destroy the Communist forces and to arrest their leaders. The rugged, war-hardened Eighth Route Army, together with its Party leaders and devoted followers, escaped annihilation by slipping through a gap in the encircling forces and marching a long and tortuous route into northern China. Eventually they gained a foothold in Yenan, a remote locale where they could regroup and rebuild their strength. The Communist Party remembers this event as the "Long March." During this arduous march, casualties were high, but the Party won the hearts and minds of the less privileged along the way. The glorious ideal to save China and to liberate the people from tyranny was strengthened.

By 1948, only two years after the Government force's triumphant march along Nanking Road in Shanghai—in celebration of the final victory against the Japanese—the country was in a critical state. The destruction of resources during the civil war and the inattention to the rebuilding of the tattered country had bankrupted China. Super-inflation, on par with German inflation after the First World War, shocked the people and the Government. All signs indicated that Chang Kai Shek's Nationalist Government would fall at any moment unless his elite military forces could win the battle against the Communists. I sang verses from the Eighth Route Army marching song with naïve gusto.

> Chinese people have reached the most dangerous time.
> We are compelled to utter our last roaring conviction.
> We, the millions of people, must unite.
> Have only the conviction, to march forward.
> Under the enemy's heavy shelling from the guns,
> March forward, march on forward."

I was oblivious to the fact that I was singing a Communist Army marching song in a Nationalist-controlled area. We had no idea that this song would later be chosen as the new National Anthem of the People's Republic of China in 1949. At that time I was too young to realize the gravity of the situation. The only thing that caught my interest was the abundance of new bank notes issued by the Central bank in an attempt to deal with inflation. The bank notes were artfully printed and bore denominations in tens and hundreds of thousands of yuans. I chose to hold on to bank notes with certain types of serial numbers. After over 60 years I still have a wad of notes with symmetrical serial numbers—such as 2345432 or 2222222. I continue to find numbers fascinating.

Our concern over escaping from Shanghai to Chongqing returned, only this time we were considering leaving Shanghai for good. This was almost unthinkable for all of us, especially for my father who was the one who actually had to make the difficult decision. He had a successful legal career and was very settled in Shanghai. But when the monumental battle between Chiang Kai Shek's elite forces and the battle-hardened forces of Mao saw the collapse of the Nationalist army just north of the Yangtze River, my father decided it was time to abandon his hometown for Hong Kong. During the final few weeks, the number of my classmates dwindled as more families decided to leave Shanghai. We all were busy asking each other to write things down in our autograph books before we went our separate ways. We all vowed to meet again sometime, somewhere, hopefully back in Shanghai.

En route to Hong Kong we stopped in Taiwan to visit some of my father's relatives and to consider Taiwan as a possible new home. Taiwan turned out not to be an ideal fit, so we continued our journey to Hong Kong. My parents were concerned about starting their lives over without their previous high social status, and their two sons were preoccupied with adapting to an entirely different education system that conducted classes entirely in English.

Hong Kong in 1949, relative to Shanghai, was a small and tranquil city. My first impression from the ship was its striking beauty. The Island of Hong Kong had been ceded to Great Britain in 1897 as compensation for the loss of the notorious opium trade. Behind the well-developed city stood the mountains, green with leafy trees and vegetation. Victoria Peak was the tallest spot on Hong Kong Island, while Lion Rock Mountain was the backdrop for Kowloon. Mansions framed the shoreline. The Hong Kong Shanghai Bank Building and the Bank of China Building dominated the skyline of the Island, while the colonial-styled Peninsula Hotel and the clock tower of the railway station dominated Kowloon. The harbor bustled with boat traffic.

The British Navy maintained its presence through a naval base in the harbor. A common sight for cross-harbor ferry passengers were the visiting frigates and battleships of the Fleet with their colorful bunting and signal flags hung out in cheerful announcements of their arrival. The sailors from these ships made Wanchai their local hangout, giving the area its spicy and sexy reputation. Films and novels set their tales of murder, love, and daring adventures in these neighborhoods, which embellished this reputation.

You can still find plenty of exotic dance clubs in Wanchai, but with the diminishing number of sailors in this hi-tech modern world, business has steadily declined. Hardware and construction material stores have taken over. Wanchai, now rundown and with most of its buildings dating from the beginning of the last century, is slated for redevelopment.

The boat bringing us from Taiwan docked at the Ocean Terminal, at Tsim Sha Tsui on the Kowloon side. As I disembarked with my family, I found myself assailed on all sides by a babble of nonsense sounds! Cantonese was incomprehensible to me, and since I could not talk with anybody in Mandarin, I resorted to English. Fortunately, English was the *lingua franca* for the educated of Hong Kong but the general population did not understand English. I had found myself in an alien land.

Many other families from Shanghai had made their way to Hong Kong before us, so there was a thriving Shanghainese community that had settled mainly around the North Point district. Here I found shopkeepers who understood me, and my parents made new friends. My parents never did learn Cantonese; my younger brother was probably the most successful language learner. My second maternal uncle and his wife had lived in Hong Kong for some ten years before our arrival and they still had not mastered the local tongue. My four cousins, all about my age, had the advantage of being exposed to it from birth, as did the three sons of my other maternal uncle.

I adapted easily to the education system. Every subject, other than Chinese and Chinese literature, was taught in English, which helped to prove that once a nonregimented learning skill is acquired, the progress of acquiring new skills and knowledge is made easier and more natural. I was on my way to becoming multicultural. I appreciated the discipline of reciting catechism and passages from the Bible at St. Joseph's College, the Catholic school I joined. Reciting biblical passages allowed me to quote freely in and out of context to illustrate new meanings. This turned out to be a conundrum for me, as I could convince myself that the Bible was unreliable or I could conclude that the explanations provided by the priests and other learned people were wrong. This discovery began to shake my belief in the Bible.

My five years in Catholic school were an important revelation. I was deeply influenced by the sincerity of the Jesuits and I began to view religion as my own personal confessional. This is something akin to believing Confucius's instruction that everyone should review one's daily actions. For example, "When dealing with friends are you being faithful?" This is an easy way to cleanse one's soul. However, I was never comfortable with religious dogma. With our limited ability to use words to express ourselves, we can never be sure that we have the ability to accurately communicate what we want to convey. We must always be aware that we can never clearly understand who we are, why we exist, and how we conjure up the means to reconcile our fictive beliefs.

As I progressed in school, I faced the challenge of sitting for the School Certificate Examination. This is meant to be an exit examination for the secondary school student. Each student can choose to sit for up to ten subjects. I bravely took all ten subjects.

At that stage of my learning, I found all science and mathematics easy, so I focused my preparation by cramming dates and places that were important in relationship to historical and geographical events. I found the way history and geography were taught made these subjects uninteresting and difficult to remember. When the examination results came out, I was elated. I was among the top ten students in Hong Kong and received a scholarship.

Those of us with high enough grades were given special privileges. At the time we did not fully appreciate that the privileges were designed to be one way of teaching a sense of responsibility. We were given tasks as class monitors and teacher's assistants, marking exams. In our courses we were guided to find the answers ourselves instead of merely absorbing what the teachers dictated. It was a wonderful way to acquaint us with self-learning through the use of library materials and through discussions with the teachers at tutorial time. On reflection, this was a better method than the system that had just been introduced in the UK at that time. The British system abolished the matriculation year but introduced a two-year Advanced Level Study, calling for students aiming to enter universities to study a minimum of three Advanced Level subjects that matched the student's specialization. For example, for Engineering, the students must choose three from a list of subjects—pure mathematics, applied mathematics, physics, and chemistry. These were regarded as the basic courses that could be taught outside the university environment, and that would also serve as the needed tools for dealing with engineering subjects. The real aim was to lower the cost of the university education by reducing the number of university courses from four years to three.

With my multiplicity of interests, I decided to study electrical engineering. Even though I was fascinated by chemistry, for some inexplicable reason I decided that I wanted to be an electrical engineer. As it turned out, the Engineering Faculty of the University of Hong Kong could not get the Electrical Engineering Department equipped and staffed for students that year, and my only recourse was to apply to a university in Britain. I chose Woolwich Polytechnic, which offered the one-year pre-university course I needed in Britain to be accepted into the new three-year degree course. Woolwich Polytechnic was one of the many internal Colleges of the University of London. I could study for my A-level subjects first and then continue there for my degree courses. Because of my interest in Chemistry I took four subjects for the A-level examination, even though the university requirement was three. I chose A-level chemistry because I wanted to have the opportunity to play with different experiments. I did not consider science and mathematics to be work.

With all of this enthusiasm and naïveté, I looked forward to the summer of 1953, when I would board a ship for London. Even though I had never been away from home, I was cool, calm, and collected throughout the long summer, while my mother busied herself putting together everything she thought I would need for my trip to Great Britain. My mother viewed London, which was the second largest city in the world, as a deserted village with nothing to eat or to buy. To her, foreign lands were barbaric and uncivilized. So she proceeded to pack a large trunk with enough clothes, toiletries, canned foods, and numerous other items for the next ten years. Her furious packing caused me to recall a famous Tang Dynasty poem that expresses a mother's concern for her departing son: "She knitted the sweater with care, stitch by stitch, till it was well made and complete. For the son who is departing for long." It concludes, "Who truly knows a son's deep appreciation for the kindness of a mother!"

My mother chose to remain at home while Father and Timothy went to see me off. My mother was the pillar of strength in our family. She always had the self-discipline to control her emotions and could make the difficult decisions with accurate judgment. She realized the importance of her two sons' receiving an education, and she hoped that we, in turn, would succeed in reaching financial independence. Sending the elder son abroad to an expensive school would drain their resources, but it needed to be done. There would be no spare money for me to visit home over the next several years. In fact, it would be exactly thirteen years before I returned, with wife and kids in tow. So on that summer day in 1953, she decided to bid goodbye from within her home instead of waving goodbye from the dock. She did not feel it would have been proper for her to become emotional in public.

Many relatives came along, as they were all eager to have a look at the S.S. Canton, which had plied the UK-to-HK route for a number of years. The thickness of the white paint can be used to measure the age of the ship, we were told by one sailor. He explained that the ship was repainted each time it was readied for a long voyage. It did look spick and span, and ready for the thirty-day journey. When the announcement came over the public address system instructing all visitors to leave the ship, I was still conducting a ship tour. Up until that point, I had taken everything in stride, chatting with my relatives and friends, remarking on the cleanliness of the deck, reminding everyone to write me, urging them to move more quickly so that they wouldn't miss the gangway. Spotting a comfortable chair nearby, I sat down for a moment to gain my bearings. At that moment I finally realized that I was severing, once and for all, the umbilical cord. I rose unsteadily from the seat and walked towards the railing that faced the dock. The band was playing, people were waving, the gap between the ship and the dock growing ever wider. I was leaving, completely by myself, bound for a far-off land. I was only nineteen years old.

My cabin was a standard four-berther located a couple of levels below the main deck. As I approached the cabin, three men, all of whom looked old enough to be my uncles, greeted me warmly. It seemed natural that they would be there to take care of me. One of them suggested that I request an upper berth even though I had already been assigned to the lower. He said in a persuasive tone, "You are young and agile, you will find the upper berth ideal for the trip." I was delighted by the idea. The three men fit perfectly the Confucian saying, "When three people approach you, one of them must be your teacher." Later during the voyage, Dr. Tsou, a lecturer from United College in Hong Kong, taught me the rudiments of tensor mechanics. He became a guiding hand in a number of arenas, including helping me to appreciate the tastes of hot chili and curries, to discover the secrets of beer as a way of suppressing the intense heat of the food, and how to enjoy learning higher mathematics significantly beyond my understanding at that time. Years later Dr. Tsou would become the head of the Mathematics Department at The Chinese University of Hong Kong. He was still teaching there, though nearing retirement, when I was recruited to lead the University in 1987. He exemplified the type of teacher who devoted his life to inspiring and motivating both students and colleagues, and one of his students was even awarded a Field Medal. My voyage on the high seas was off to an auspicious start.

The wind gathered speed as we reached the open sea en route to the first port of call, Singapore. I had little knowledge of Singapore, apart from the captain's announcement that the journey would take four days. As he signed off with that first announcement, every sway of the ship ploughing the waves caused my stomach to churn. I did my best to focus my gaze towards the far horizon for a greater sense of steadiness, and moved away from the side of the ship where the diesel smoke was belching out of the engine. It was all to no avail. My stomach heaved, a cold sweat poured down, and I felt my face turn green. For the next several hours, the only place I felt more or less comfortable was in bed. The next morning I felt perfectly fine, either because the seas were calmer or because my body had gradually adjusted to the swaying. I paid less attention to the ship's motion and found myself growing an excellent pair of sea legs. The many novel activities—deck tennis, tombola sessions, eggs and bacon and sausages for breakfast, mid morning tea and cakes—became my focus. When we disembarked at Singapore, it momentarily felt like Singapore was experiencing an earthquake. It was, of course, an illusion caused by my brain retaining the control algorithm for coping with the motion of the ship. It should have dawned on me then that motion

sickness is the result of attempting to counter the motion instead of swaying with the motion.

My very first visit to Singapore left hardly an impression on me. After
nearly a week at sea, the attention of the passengers was principally directed at getting off the ship to stretch our legs. Dr. Tsou, my mentor and
the self-appointed leader of a couple of other teenagers en route to boarding schools in the UK, ushered us down the plank to the unassuming
dock of Singapore. If it had not been for Dr. Tsou, I probably would have
stayed on the ship eyeing the low-rise buildings along the dock or gazing
at the green backdrop of various palm trees, too nervous to venture off
the ship into strange territory.

From Singapore the S.S. Canton sailed westward across the Indian
Ocean towards the tip of the Indian peninsula where Colombo, the capital of the island state of Ceylon, now known as Sri Lanka, is located. The
ship made a short stop; long enough to allow us to disembark to lounge
on a colonial-style promenade in luxurious splendor, sipping cool fruit
juice under a line of magnificent palm trees. The next stop was Bombay,
now known as Mumbai, and the famous Hanging Gardens. I never made
it to the Gardens, nor did I even enquire whether the ancient Gardens
still existed, though I did stroll through the city's broad avenues filled
with all types of vehicles generating noise and dust. As we sailed along
the Red Sea, our ship gave me the impression that the heat had slowed
everything to a standstill. Mile after mile of flat, featureless sandy shores
ran together as the ship glided smoothly towards the African port of
Suez. There were no waves, no breeze, but plenty of heat. Life was a dozy
routine. Half awake, we ate our breakfast, morning tea, lunch, afternoon
tea, dinner, and finally supper, if anyone still had an appetite. The days
passed quickly as we lazed about.

As we neared the Suez Canal some of us got very excited about the
opportunity to watch a massive ship rise and lower in the lock, which
was just large enough to accommodate our ship. For the first time in my
life, I was getting to see an operating lock. At the same time, I was keenly
aware that the lands along the two shores were the ancient civilizations
of Egypt, Syria, and Jordan.

Our stop at Port Said was too short to make a tour of the pyramids or any of the famous relics of Egypt, though I did disembark from the ship so that I could boast that when I was still a teenager I had already been to Asia, Africa, and Europe.

Our journey through the Mediterranean offered us a glimpse of the northern coast of Africa. We could not discern individual cities but according to the map we passed by Tangier and Casablanca. When the ship approached Gibraltar, all of the passengers were on deck to view the twin peaks on either side of the narrow passage. They looked like a battleship with twin control towers guarding the strait.

The transition from warmth to cold came upon us unexpectedly as the ship steamed into the English Channel. The wind howled and the rain pelted down. The enormous steamer was quickly shrunk to an insignificant toy battling mountainous waves. We felt like true mariners, as in less than an hour the sunshine of the Mediterranean had been completely blocked by thick, gray clouds and ominous, threatening weather. The sea eventually calmed and our arrival in England was a typically gray and dull morning. There were no bands of colorfully attired musicians saluting the arrival of passengers and the crew of the S.S. Canton. The ship simply slid into the Port of London at Tilbury, a small town on the estuary of the Thames.

Immigration officers sat behind a long row of tables, scrutinizing the passports and various other documents of the disembarking passengers, as if all were of dubious character, and intent on crashing our way into Britain with nefarious plans. The process was played out over several hours. By the time I got to the head of the line, I was completely exhausted. The immigration officer quickly looked at my documents and wished me good luck with my studies.

The next anxiety was finding the British Council representative, though this turned out to be a short-lived fear, as I quickly noticed some of my young friends from the ship already waiting in a group under the British Council banner. Our group was told to board the train going to Victoria Station in London. There we would meet other British Council representatives and collect our luggage.

The train had separate compartments in each carriage. I found myself in one of these compartments along with seven other people, all strangers. En route to London, the scenery of green fields was marred by rows of drab tenement houses. I remember distinctly a remark made by one of the train passengers. "Just look at those houses—lower-class people live there. The houses in our colonies are far superior. We live in spacious palaces with servants to attend to all our needs. I cannot understand why these Brits so lack the will to succeed."

As we got off the train, I never imagined that England would become my adopted home for the next decade and a half. I was greeted immediately by people from the British Council, an organization responsible for developing relationships with foreign countries, especially those within the British colonies. We were directed to wait together so that we could be transported by bus to Lancaster House, a mansion that served as a temporary boarding house for new arrivals. Here we were to receive orientation before being set free to fend for ourselves at different locations, mostly at university dormitories or as guest boarders in private homes.

The first thing that struck me about Lancaster House was the sheer size of the hall. It had high ceilings, wood-paneled walls, and wood flooring, all stained the same dark brown. I could imagine it being a banquet hall of a medieval castle except that the chandeliers had light bulbs and there were no rich brocade curtains. The floor was bare of carpets and furniture, save for a single row of chairs placed around the four walls and a few long tables. All of our activities were held in this rather forbidding hall. After two days I discovered that Lancaster House was near Hyde Park. After a short walk I was in a great expanse of lawns dotted with magnificent trees. It was such a contrast to the concrete jungle of the urban areas surrounding it, and it was my first real taste of "That emerald isle."

Isolation was the best way to describe my first few days in London. Among the millions of people in the metropolitan district, I was one tiny ant. It took me well over an hour on bus to travel from Central London to the borough of Woolwich and beyond that to reach Wrottesley Road in Plumstead, one of London's outer boroughs. Luckily I did not have to carry my gigantic trunk or other suitcase. The luggage had already been delivered courtesy of the British Council while I was undergoing my orientation at Lancaster House.

Wrottesley Road was a tree-lined street off the main bus route. No. 14 was the fourth semidetached house on the right. It was a quiet suburban street leading from the main road that sloped gently up to a smaller road. As I walked from the bus stop, I caught sight of a handful of faces peering out through windows. The entire area was residential, quiet and deserted, save for the rare pedestrian, bicycle, or car. I felt like the last soul on earth, approaching a strange house with nothing but my backpack.

TAKING A PICNIC BREAK DURING A CARTRIP WITH CLASSMATES IN ENGLAND

Building Independence

"This is independence," I said to myself, as the odd-looking red double-decker bus, bearing the route number 53 and final destination Plumstead, carried me from central London towards my future lodging. The agent had already sold me a ticket, a white receipt duly produced from his ticket printer. He punched some buttons on this ticket machine, turned a small handle on the side of the printer, and out came the ticket. It was my first ride on this type of bus, which would become my daily form of transport in London for quite a few years.

I loved those tickets, each with a four-digit sequence of numbers. During that time I took the number 53 bus from Plumstead on a daily basis to the Woolwich Polytechnic where I was studying. The journey only took around five minutes. During the trips I would occupy myself by trying to factorize the ticket number to its prime components before I reached my destination. I would win if I could complete the factorization before alighting from the bus. I have forgotten all of my scores now, but it was an exciting game. I would be especially pleased if I could finish a massive prime number like 7891. I had to divide that number by 2, 3, 5, 7, 11, 13, 17, 19, 23, 29, 31, 37, 41, 43, 47, 53, 59, 61, 67, 71, 73, 79, 83, and 89 to show that 7893 is prime.

My first bus journey from Lancaster House through London passed by many famous sites. The bus sped along Oxford Street, turned at Oxford Circus, and then went down Regent Street, passing Trafalgar Square, where Nelson stands on top of the column in commemoration of his victory at Waterloo. I gazed with amazement at the grandeur that was Central London. I was too busy sightseeing to reflect on the taste of independence, a taste that I was yet to know how to acquire.

After we passed the Westminster Bridge with the Houses of Parliament on our right and the London County Hall on our left at the far end of the bridge, we entered the vast suburbia of southeast London. The famous landmark scenery was soon replaced by nondescript brick houses and shop fronts with the occasional large patches of greenery that are known as commons, as the bus traveled farther and deeper into south London.

I felt as if we had driven across the whole of England and should be nearing the coast when suddenly I noticed the name Woolwich Arsenal. Woolwich was famous for its historic role of being an arsenal where guns were produced and stored for battle. The most well-known colleges of the University of London are University College, King's College, Imperial College of Science, and Queen Mary College. These four colleges have a greater leeway in setting their curricula and their admission policies, though they do have to follow the rules set up for all colleges as defined by the Senate of the University of London. Woolwich Polytechnic is now part of Greenwich University and is still located in Woolwich. In 1953, it was not a particularly outstanding institution. As a polytechnic, it also ran non-academic courses.

The suburb of Woolwich is small. All the students studying at the Polytechnic lived on their own in rented houses, and many, like myself, in lodgings offered by landladies in hostel-like accommodations. Mrs. Cameron, a widow who lived with her unmarried son, ran my house. She took in four lodgers at any one time.

When I arrived I was assigned to a small room with simple furniture, barely large enough to accommodate my big trunk and personal belongings. When one of the lodgers, an Indian who never appeared without his turban properly and neatly in place, left, I, who by then had become the landlady's favorite, was offered his spacious room with a bay window facing the road. Most important, this meant I could work more comfortably on a large table. Laying my big drawing board on it, I could do all of my engineering drafting homework. I was so excited about finally having space to accommodate my books that I was taken in by the glossy brochures of the slick *Encyclopedia Britannica* salesman. A low-cost investment would bring me the world's knowledge, updated for the next ten years, all at one affordable monthly payment. I signed the agreement on the dotted line without realizing the financial strain this purchase would bring my parents. My belief that I would actually read the 24 volumes of this great encyclopedia, from cover to cover, and become the most knowledgeable man in the whole world proved baseless. My gullibility remained on display until the day I donated those hard-covered tomes, with the gilded characters on each spine. I ended up reading approximately 0.001 percent of the material, which was available for free in any library. The volumes traveled with me from the UK to Hong Kong to the United States. During those years any real benefit was demonstrated only a couple of times in helping my children with special school projects. After that I learned to be more wary of sales pitches. The meaning of independence was clarified somewhat, particularly when my father reminded me repeatedly that I had placed a significant financial burden on him.

Did I have a real concept of independence at that point, other than my own interpretation, which was that I could do what I liked without having to seek permission first from my parents? My buying the encyclopedia was a test of the limits of my freedom. The first hurdle must be financial independence. Or as Micawber in Charles Dickens' novel *David Copperfield* said, "Annual income twenty pounds, annual expenditure nineteen nineteen six, result happiness. Annual income twenty pounds, annual expenditure twenty pounds ought and six, result misery."

For the entire four years while I was attending Woolwich Polytechnic, my life centered on attending classes, performing experiments, socializing with classmates, and participating in sports, especially table tennis and tennis. I helped organize events from large-scale Chinese dinner parties at social clubs to dance parties at the Poly and at ULU, the location of the central students' union house. I also did community service for the Legion of Mary, visiting hospitals and assisted living centers, besides diligently going to church every Sunday. Mervyn from Trinidad was my closest school friend. He was a bright student with a boundless amount of energy and life. He said that you could tell a woman's age by looking at her heels. More mature women have a more pronounced tendon at the heel. This became a regular exercise that he conducted each and every time a woman passed by. He assured me every few weeks that his observations on women's heels were well supported by his continuing research.

Mervyn's passion for cricket ranked even higher than his interest in women's heels. He converted me into a cricket fan and into a novice cricketer. We were glued to the radio whenever John Arlott, one of the great BBC cricket commentators, was describing the ball-by-ball action in the cricket matches between the UK and Australia during test matches. John had the knack of conveying the grace, elegance, and sportsmanship of the venerable game. He would unexpectedly alert us to how neatly the spin bowler tricked the best batsman into misjudging the ball, causing the ball to touch the edge of the bat and to bounce off straight into the wicket.

Without being consciously aware, my contacts with a great variety of people increased my consciousness of human relationships. Mervyn's enthusiasm reminded me of the saying "Birds of a feather flock together." During my college days, I was developing my unconscious selection process of finding compatible persons to become my associates and friends. My search for independence was progressing in the right direction.

Tom, my landlady's son, was aloof but helpful. I seldom asked him for a favor. One day while he was working in the basement he invited me to have a look at this cavernous area. I noticed that the basement was half sunken below the ground level and had only two short windows on the wall facing the garden. If I covered them up with cardboard I could use the basement as a darkroom and print the photos that I had taken during my voyage to the UK. I had my contact print frame and a developing tank that I had brought with me from Hong Kong, but I didn't have an enlarger. My plan was to spend as little money as possible on this project, since my budget could not stand anymore frivolous spending. This ruled out the purchase of a real enlarger.

I found a large cylindrical tin with a lid just right for a light-tight housing as my illumination source: a light fitting with a light bulb could be inserted at the bottom end and a small rectangular opening, the dimension of a 35 mm film, cut open at the center of the lid, to allow light to shine through the film. Within the lid a magnifying glass was fitted so that a near-parallel beam derived through the convex lens from a white bulb would illuminate the film aperture fairly evenly. Then I borrowed a retort stand with two holders from the lab at the Poly. It served as the platform to hold the lamp and the camera in the correct positions. The camera would function as the enlargement device. Undoubtedly, this was a string-and-sealing wax contraption, but it worked the first time.

I coaxed Tom into letting me do some contact prints for him so that he would let me use the basement as a darkroom. He offered to build window shields for the two windows. In the meantime, I went to purchase the photographic papers, developers, fixer, and neutralizer. I was in business. The prints for Tom and my film development all turned out well. Everyone was delighted.

Apart from having to buy a set of plastic trays, a paper cutter, and an adjustable mask, I had spent a fraction of the cost of a real enlarger to acquire a unique home photo processing business. Admittedly, I did not have much business, but I was able to develop, print, and enlarge photos for my friends. I even got a job as the official photographer for ULU functions at the Senate House. My humble photo lab was a successful demonstration of my well-earned independence. I had overcome my lack of money and used my budding engineering skills as well.

Nineteen fifty-three was the last year that food was still being rationed, a remnant of the Second World War. At my lodging our meals were meager. I remember the hunger pangs that came only an hour or so after an early dinner at 6 p.m. I would venture out into the cold damp nights to buy fish and chips. To this day, I enjoy eating the crisp crust of the batter around the succulent codfish and the soft, vinegar-soaked, freshly fried chips out of newspaper.

Breakfast at 7 a.m. consisted of cereal and toast with bacon rashers, or even eggs, which also were still being rationed. Dinner was the only full meal of the day. Promptly at 6 p.m. we all sat around the oval table in the dining room. The table setting was laid out according to the best English tradition, forks on the left and knives to the right of the plate. A smaller-sized spoon and fork lay horizontally on the table across the top of the plate. A white serviette was on top of a side plate placed on the left of the fork. In addition, we all had a glass of water. The food, however, was much less attractive than the table setting, especially on Sundays. The Sunday dinner was called high tea. Only cold meat and bread were served. The meat was cut so thinly that light penetrated it. High tea had one high note. It was when the best desserts were served. I especially enjoyed the apple pies served with custard. Trifles and suet puddings were my other favorites.

"Necessity is the mother of invention" has resonated throughout my life. When I would tell my landlady that I needed to take a walk after Sunday high tea, I would always insist that the tea was so wonderfully hot and strong that I simply needed to digest the food by taking a walk. Actually, the wafer-thin piece of ham was so thin that it had lost its power before reaching my stomach. If I took a fast walk down the hill, I knew the golden-brown crusted fish and greasy chips would be waiting. My fellow lodgers often walked with me, in silence, until we had collected our piping hot food wrapped in yesterday's newspapers. We would open the parcel and splash the contents with salt and malt vinegar. The walk back up the steep hill seemed easier than the downhill run.

Unconsciously, I was building an increasing degree of independence. This was easy when the lodgers were all male and all students. As time passed lodgers came and went. At one time we had a teaching member of the Polytechnic and a gas company engineer, whose hobby was collecting butterflies. Then one day two young ladies replaced the two working-men, which transformed us into a mixed-gender hostel. The presence of females changed the etiquette within the house. It also broadened my horizons as it was an experience with a different slant. My brother and I had no sisters and we studied in all-boys' schools. My socializing experiences with girls were next to none. This opportunity of living with two young women, Kathleen and Mary, was new to me, and initially caused me to be very much ill at ease. Coming directly from Ireland to start their working careers, they were just as much strangers in a foreign land. Despite the closeness of the living arrangements, the relationships actually developed as if we were brothers and sisters. Kathleen and Mary were both Irish Catholics and I, under the influence of Mervyn, had become Catholic. We often all went to Mass together on Sundays.

Mervyn and I, in our non-class time, worked for the Catholic Society of the Polytechnic, as well as the Legion of Mary. At Society functions we organized theological debates and discussions. On Flag Days I stood on street corners, stopping passers-by to make my sales pitch for whichever charity was the flavor of the day.

Kathleen and Mary occasionally came to our activities at the Poly and at the church, so I had a fair amount of social contact with them. A year or so later, Kathleen was engaged to be married. She asked Mary to be her bridesmaid and, to my surprise, asked me to be her best man. They were married at a Catholic Church on a cold and windy Boxing Day. I nearly froze without my overcoat as we posed for pictures. I wondered how the bride felt in her sleeveless wedding gown.

A stream of correspondence with my younger brother and Father kept me apprised of the situation in Hong Kong. Timothy progressed well at school and later entered Hong Kong University. My father repeatedly did his best to urge me to do my utmost, particularly in my studies. He cautioned me about money often, even though I had been very responsible after the encyclopedia incident. I felt under no pressure, and life was rich with opportunities for gaining knowledge and experience. Time passed quickly and it was soon final examination time for my degree in electrical engineering.

My college years were filled with activities, most of which seemed entirely extra-curricular. During my high school days I found science subjects fascinating and easy to score high marks in without having to labor over them through sheer memory work. The meaningless dates and events and places as in history and geography were less fun. I carried this same attitude into my study of science subjects at the college level. I never missed class, and I could complete the assignments so quickly and effectively that I had plenty of time for non-academic work. I seemed to be attracted to organizational work. I joined a number of student societies, including the Catholic Students Society and the Chinese Students Society. Tennis eventually consumed most of the time that I should have been studying. The weather was beautiful the summer of my finals year and very conducive for outdoor activities. I played tennis almost daily, even on the day before the exams. Perhaps I was a bit overconfident, as I missed getting a First Class Honor for my degree. I did manage to get a Second Class Honor instead of merely passing.

"Regret" is a word that I seldom use to describe a missed opportunity. When I finished my first year, I was supposed to apply to a university of my choice. My enthusiasm for science was so overwhelming that I had insisted on taking four subjects: mathematics, applied mathematics, physics, and chemistry for my advanced level examination. I passed these easily and, therefore, I had a great opportunity to get into any of the more renowned Colleges in the London University system. With four A-level subjects, I should have been able to gain acceptance into the more famous colleges without much problem. However, I did not realize that I needed to initiate an application and instead by default continued to attend Woolwich Poly. I do not regret missing the opportunity. I found that the closer relationship with the teaching faculty members permitted a higher level of direct interaction with the teachers.

My graduation from university was a grand affair held at the Royal Albert Hall. The Queen Mother, who was the chancellor of the entire University of London, presided at the long ceremony in which all of the graduates of 1957 marched past her, each stopping momentarily to doff their hats before filing off the stage. Solemnity, pomp, and circumstance were the correct descriptions for that occasion. Every student was photographed individually the day before so that they would have the picture as a memento of the occasion. I was not aware how trying it was for the Queen Mother to smile graciously to each and every graduate, numbering in the thousands, without slipping into a yawn or, even worse, a frown. When I, as the vice chancellor of The Chinese University of Hong Kong, had the experience of being in that same position at our own graduation ceremony, I realized then how enormous a task it was.

In 1957, receiving my Bachelor of Engineering was definitely a highlight. I had finally reached that stage in my life where I needed to earn a living. Years earlier I wrote in one of my letters home to my father that as soon as I graduated I would be financially independent and that he would no longer need to send me money. The time had finally arrived. Earlier in the year, I had prepared my résumé and followed my teachers' advice on how to meet the recruiters from various companies coming to the Polytechnic to interview job candidates. By the time I went to the graduation ceremony, I was actually already working in the company of my choice, Standard Telephones and Cables Ltd (STC). I was interested in telecommunications, and STC was one of the best companies in the industry in UK. Besides being a well-known company with a long history, it offered a graduate apprentice program. I would be assigned to four areas in rotation and then would be able to choose the area I liked most. The salary was eleven pounds sterling a week.

During my college days, I had the opportunity to work during the summer vacation as a student apprentice. This turned out to be a great way to introduce students to industrial life. My first assignment was to Metropolitan Vickers, a heavy engineering company that made electrical generators and transformers, located in Manchester. As a student apprentice, I lived with a family and was well looked-after. It was a great first experience as a working person.

The work, however, was simple and at times boring. We student apprentices were attached to work units responsible for the production of heavy electrical equipment. We moved from unit to unit each week. We were given trivial tasks, such as filing the burrs off the edges of a newly cast rotor for hydroelectric generators. The rotors were huge. I was given a ladder and a file and was told to do what I could. It was a new and fun experience because I could see the immediate result of my efforts. I could sit on top of these rotors, ten or more feet in height overlooking a vast working area, to watch the activities around me and to imagine myself actually contributing to a production effort.

The winding of transformers was the most interesting part of the six weeks of training, and the reward of actually being paid for acquiring the experience. When I was playing with the small transformers in my radio set, the primary and secondary windings made use of thin wires. For a large power transformer, the copper wires were very thick so that a large current could be transmitted without causing the transformer to overheat due to excessive ohmic loss. These wires are rectangular in cross-section, around 3x4 mm, for ease of winding in layers. The secondary wires had many turns and a smaller cross-section. Every so often the winding machine would stop to allow the empty spool to be replaced with a new one. The new wire must be carefully welded and shaped correctly before the process continued. When both primary and secondary wires were in place, the transformer core was complete and it was raised away by a lifting device for it to be readied for its initial testing. I was able to follow such an operation from start to finish and learned a lot about the differences between theory and practice. The importance of understanding the inadequacies of a technology handbook when it is to be used as the sole means for a technology transfer cannot be overemphasized. Throughout my life as an engineer, at every stage of my career, I reminded myself that a range of well-trained and experienced personnel are indispensable in the course of work on any engineering project. My first and subsequent apprenticeships were very valuable.

A week or two later, after I returned to the Polytechnic, I received a letter from one of the vacation trainees with whom I had become friendly during our work with Metropolitan Vickers. She invited me to visit her home in London and gave me detailed instructions on how to get there using the underground Tube system. From Woolwich to North London I had to take a train first to Charing Cross on the Southern Railway Line and then the Tube to where she lived. It was a beautiful sunny day when I went to meet her. Her parents welcomed me into a large detached house with a well-tended front garden. After some pleasantries, they said that Doris would look after me and that they would be going out. Doris suggested that we should enjoy the great weather by taking a walk in the neighborhood before returning for lunch. She lived near the northern edge of Greater London. It was a high-class residential area near a Tube station, as well as the vast open spaces of farmland and wooded areas. Doris took me on a long hike through picturesque paths. We met very few people and those we did meet all seemed to be nodding acquaintances of hers. I almost felt like a trophy new boyfriend whom she was showing off.

Lunch was not on the table when we got back, but she had organized it to be easily and stylishly presented almost instantly. Compared with my lodging, this was palatial. A soft tablecloth with embroidered edges, stiff white matching serviettes, shining cutlery, and fine bone china crockery defined the ambiance. I was impressed with her food and her ability to put me at ease. I was only worried about what would come next. I was not sure whether I could handle the situation with her parents out and us alone. What would her next move be? Would I be able to think straight and deal with it appropriately? Should I be a wimp and start to make excuses about having to make the long trek home? It turned out I was a wimp, and I could sense that Doris was not very happy as I was leaving, thanking her profusely for her hospitality.

My second vacation apprenticeship was with GEC at Coventry. Being a second-year student and about to enter my final academic year, the training involved testing electronic equipment and carrying out experiments. The pay was higher and I was beginning to feel more like a wage earner.

It was not without some trepidation that I put pen to paper and announced to my father that his son was now financially independent. The timing of this event revolved around me receiving a confirmation letter that I had been offered a post as a graduate engineer at STC. My first target for gaining independence had been achieved. I made an estimated balance sheet several times to make sure that the balance I had at the time of writing the letter would see me through to when I would receive my first paycheck. I was sure that my father would breathe a sigh of relief that he had managed to support his son's education, a task he promised himself that he would manage. I had not taken into consideration that I was no longer an hourly paid employee who received a weekly paycheck. I was now to be paid on a monthly basis. Not knowing that I could have asked the company for an advance, I had to humbly write to my father for help with food money until the end of that long first month.

WORKGROUP AT STANDARD TELEPHONES AND CABLES, WOOLWICH, UK (CA. 1960)

Hitting the Jackpot, Part 1

"There is a tide in the affairs of men, which, taken at the flood, leads on to fortune; omitted, all the voyage of their life is bound in shallows and in misery." I learned this beautifully crafted sentence of Shakespeare by heart when I was studying *Julius Cæsar* at school. It refers to the notion that timing is everything in achieving a goal. It is also a reminder that all the necessary components must come together before any major changes can occur. I entered the workforce when the demand for telecommunication systems and hardware was at a crescendo. This particular tide was created by a strong public demand for broadband information transmission that could relieve the congestion in telephone traffic, as well as allow video signals to be transported efficiently via cable. How far would this tide carry me along?

In 1960, when I joined the research laboratory, the research community was already eagerly responding to the public demand for improved telecommunications. A large sector of the work at Standard Telecommunications Laboratories Ltd. (STL) was devoted to improving the capabilities of the existing communication infrastructure. The major focus was the research and development of a transmission system based on the use of microwaves at a millimeter wavelength.

During the few years that I was working at Standard Telephones and Cables in Woolwich, I was engaged in the development of the next-generation equipment for the market. The equipment specifications were already well defined. As a development engineer, I was constructing the parts to a new system that would be able to convey information at a capacity of 50 percent more than the previous system. This had been achieved by using a radio wave at a frequency of 6 GHz instead of a radio wave at

4 GHz. We were making a replica of a known system, but the equipment now had to handle 50 percent more information, while the wave-guide components were scaled to fit the frequency change from 4 to 6 GHz.

The millimeter wave system was a giant leap into uncharted territories. First, the radio wave carrier was chosen at a frequency of 35 GHz, and later 70 GHz or higher. This meant that the carrying capacity could be more than ten times larger. The challenges were enormous since radio waves at such frequencies could not be beamed over long distances due to the absorption of radio wave energy by the atmosphere. The wave required a wave-guide. For low transmission loss the wave-guide needed to be made by spirally winding a copper wire on a cylindrical mandrel and then strengthening the wire by curing an epoxy coating over it before withdrawing and removing the mandrel. The wave-guide was made section by section and joined to form a straight wave-guide. Such a wave-guide would allow a particular wave-guide mode called HE11 mode to propagate with very low loss over long distances. This wave-guide system was called the circular wave-guide system, or the long haul wave-guide. Despite the significant technical difficulties of manufacturing the wave-guide and the linear requirements, R&D work started in the 1950s had reached the deployment stage for a trial system by the early 1960s.

The circular wave-guide development was initiated at a time when telephone companies had a monopoly on the market. The telephone company generally controlled the network, the service provider, and the equipment supplier. Thus the development of a circular wave-guide could be economically managed. When antitrust laws went into effect, and especially when alternative and improved communication systems were developed, the competitive pressure created a higher standard of living at a much lower cost. Optical fiber communication and new technology advances altered the fundamental economic basis.

As a greenhorn in the R&D world, with practical product development experience at STC, I lamented the slow pace and fundamental limitations of improving our microwave point-to-point transmission systems. After three years, I believed I had a thorough understanding of the current technology and was ready to change jobs.

I applied for a position as a lecturer at an institution in the UK Midlands. By that time I was married but without the added responsibility of children. After receiving a letter of appointment following an interview with the head of the Electrical Department, we traveled north to explore the housing situation. We found a new development being built of small detached homes in a nice area and we were persuaded to put down a deposit of £25 to secure a future purchase. That sum, seemingly minute now, was a considerable fortune for us then. Returning after our weekend jaunt, I gave my required one-month notice for quitting my job at STC.

The repercussions of that notice were unexpected. My supervisor called me in to his office and said that the company did not want to lose my services. The director of STL, the research arm of the company in Harlow, Dr. King, persuaded me that I should pursue research instead of an academic career. He said that I could join the research on optical communication that was happening in the Labs. With the invention of the LASER in 1959, an opportune time arrived to consider working on optical communication with a view towards creating a potentially infinite bandwidth network for the world. Dr. King emphasized the point that the project was being directed by the brilliant inventor Alec Reeves, who had invented the PCM system for digital communications in 1936 at the sister research lab of STL. Under Reeves, an optical communication project had already been launched a couple of years earlier before the birth of the laser. The light source was to be a hollow cathode plasma light emitter and an internally silvered duct was to be used for guiding the light. All of these projects were in progress when I joined STL in 1960.

When this counteroffer came, I was in a quandary. At the time I was more worried about how to get my £25 back! If I reneged now on the academic position, the institution would be quite angry. STL assured me that their lawyers would be able to get back my deposit and release us from the purchase of the house in Loughborough. They would smooth over the problem with my would-be new employer and compensate them for the expenses they had incurred. STL even offered to find a job for Gwen at the Labs! It was an arrangement I could not turn down. So I found myself looking for housing all over again, but in a different town.

Harlow was a new town built to a master plan to house the overspill from London. Much of the war rubble in London had not yet been cleared away by the late 1940s when the first homes were built in Harlow. People were desperate for homes and willing to move out to the countryside. The original plan was for the community to live and work in the town. Light industries were moved into the area. With this plan in mind, minimal car parking facilities were built and the streets were laid out in attractive narrow winding lanes. It rapidly became a town for young families. But people eventually commuted out of the town for work. It is not feasible to try to confine the radius of travel. In Harlow the cars were squeezed along the narrow streets, causing numerous obstacles that had to be negotiated by every passing vehicle.

When I moved there in 1960, the town still looked sparkling new. The town center was a traffic-free shopping area. There were bicycle lanes and walking paths to all the facilities needed to make life pleasant. It was a big change to be in the countryside with the fresh air and fields. We both had until then lived only in major metropolises. We were offered a small townhouse for rent by the town council. The house was in a row of homes that was bordered by a field of green. Brambles of blackberry and wild hazelnut were part of the hedge groves. My commute to the Labs was just ten minutes.

To start with I was assigned to the long haul wave-guide group under Dr. A. E. Karbowiak. On the first day, he showed me the actual circular wave-guide. It was a great excitement to see the specimen and to hold it in my hands. It was an engineering marvel. He suggested an assignment of looking for new transmission methods for microwave and optical transmission. He said that a clear understanding of ray optics and their relationship with wave-guide theory did not exist, and that I should look into approaches that use both ray optics and wave theory to gain a better understanding of wave-guide problems. The wave propagation within an oversized wave-guide could be expressed by both wave theory and ray optics. It was a good beginning.

Some time later, my boss encouraged me to pursue a Ph.D. while carrying out experimental work at STL. Professor Barlow, a pioneer in microwave technology at the University College of London, later supervised my work. I chose "Quasi-Optical Wave-guides" as my project. I was interested in better understanding free space propagation and guided wave propagation in closed and open wave-guides: For the closed wave-guide, the wave is bounded by a closed metallic shield, for the open wave-guide, the wave is guided by a dielectric material such that the electromagnetic field is not confined by metallic boundaries. The specific investigation was performed on an overmoded rectangular wave-guide with a right-angle bend, using a quasi-optical approach. I was delighted with this arrangement. The assignment allowed me sufficient scope to explore both microwave and optical waves for communication purposes.

The overmoded rectangular right-angle bend can be analyzed by tracing a ray using a plane wave representation. At the same time I could postulate an aperture field distribution and then construct cylindrical wavelets illuminated on the 45-degree reflecting surface. I could load the reflector portion by adding shaped dielectric materials that restored the wave front and reduced the higher order mode generation or increased the higher order mode generation. Some semi-quantitative results and theoretical analyses were made. While an exact theory could not be established, the project gave me a crude way to estimate the diffraction pattern that takes place within the right-angle bend. The theoretical method to merge the plane wave limit and diffraction limit cannot be formulated. This is still an unsolved problem.

In 1960, there was a great deal of expectation in the air within the telecommunications sector. The development of a long-haul transmission system, based on the use of a circular wave-guide operating at 35 GHz, was at an advanced development stage. Huge investment had already been made, and more investment was planned to move this system into the pre-production stage. With increased bandwidths, the major operating companies were promising better telephone systems and were planning to introduce a new service called picture-phone that would allow users to see and hear at the same time. AT&T was the first to announce a trial system in the 1960s and all of the developed nations took note.

At the same time, the invention of the laser in 1959 gave the telecommunications community a great boost of optimism that optical communication was just around the corner. The coherent laser was to be the new information carrier with a carrying capacity a million times larger than the point-to-point microwave information carrier. This theoretical performance prediction was deduced from the fact that the optical carrier frequency was at $3\times10E+15$ Hz while the microwave carrier was at $3\times10E+9$ Hz. Hence, the optical carrier should be able to carry a million times more information for each optical carrier. A race between the circular wave-guide and the optical communication was in the making, but the odds in favor of the circular wave-guide were overwhelming. Optical lasers in 1960 were at an early stage of development. They were experimental devices that had just been demonstrated at one or two research laboratories in the world. Their performance characteristics were far from those needed for use as the carriers of information for long-distance transmission. The race appeared to be a nonstarter for optical communication.

How could we so readily dismiss the laser as a nonstarter? Why weren't we able to find a wave-guide for optical waves? The concept of optical communication was simply too good to be left on the theoretical shelf.

I asked myself a series of obvious questions to see whether I could come up with any answers. Two of the questions I posed were:

1. Is the Ruby Laser suitable as an optical carrier for communication applications?
2. What material has a high enough transparency so that light at wavelengths is compatible with achievable lasers, and can travel through it over long distances?

If no answers could be found within a couple years, then perhaps optical communication would be scrapped. But the potential benefit was so large that it would have been a waste not to put the effort into exploring it in depth.

Fortunately for me, my management was happy to give me the opportunity. I was riding the right tide, a tide created by the monopolistic telecommunications industry that believed in the need for increased transmission bandwidths for making broadband communication a practical reality. The industry strategists argued for the addition of new services to expand the revenue base through increased telephone demand, as well as the possibility of introducing new services such as picturephones. The drive for making broadband communication a reality was on. When the invention of the laser hit the news, optical communication became a real possibility, since it could theoretically provide even more bandwidth for the communication systems. The attraction, especially as perceived by the monopolistic communications industry, was so great that all speculative approaches to make broadband communication a reality were supported with generous R&D monies.

At that time there were only two groups in the world starting to look at the transmission aspect of optical communication, and a number of other groups that were working on solid-state and semiconductor lasers. The lasers emit coherent radiation at optical frequencies. Their invention was a fundamental breakthrough in physics that opened the way for studying coherent optics. So the work on lasers was regarded as of great importance in many fields and was not specifically directed to the communications application. On the other hand, using coherent radiation from a laser as a communications carrier appeared to be a difficult, if not impossible, proposition. I started by following up two aspects of optical transmission systems. One was to define the ultimate benefit of optical communication. The second was to search for reasons why the eventual goals in propagation over long distances could not be achieved.

For optical communication to fulfill its promise, the light source ideally should be a single-frequency device that generates several milliwatts of power, lasts for years in continuous operation, can be modulated internally or externally, have a high power efficiency, emit radiation at a chosen wavelength, and allow the emitted light to connect two arbitrary destinations, preferably across significantly long distances at a chosen wavelength. I had to demonstrate that the probability of reaching these performance goals was finite and that demonstrations could be created to support these goals. It was a tall order, but fortune usually favors the brave.

When I moved to Hong Kong for a temporary academic appointment in 1970, the unsatisfactory state of point-to-point communication was brought home to me in a very personal way. During family celebrations such as Christmas, everyone wants to talk to at the same time with relatives in far-flung continents during the same twelve-hour period. I had to book a time slot with the telephone company, and then I was allowed just three minutes of hurried conversation before being cut off by the operator. The family waited anxiously by the phone at the appointed hour. Finally the phone rang: "This is your call to UK. You have three minutes!"

"Quick, come and say hello to your Grandma!" I tell the children. "Merry Christmas! Have you opened your presents yet? Is the turkey roasting in the oven?" "We had a great day and are nearly ready for bed." "Is it snowing in London?"

"Sorry. Your time is up!"

"Wait, wait, we haven't said our goodbyes yet. Here are the children!"

"I am sorry. I have to cut you off now."

The bill for the three minutes on the line was not cheap, either, so the phone calls were reserved for special occasions and emergencies. On top of that, all overseas calls had to go through an operator, so one was conscious of a distinct lack of privacy.

Dr. Karbowiak was a native of Poland who had immigrated to the UK, and was in charge of the research on millimetric wave transmission systems. I was in awe when I first met him. His explanations of the rudiments of the principles and the advantages of the circular wave-guide nearly converted me from an optical communication enthusiast to a circular wave-guide man. Nevertheless, he encouraged me to find a project that bridged microwave technology. "Take your time. Come back to discuss your ideas in a month." This was how he concluded our meeting.

I sat at my totally bare desk staring at its expanse and emptiness. Thoughts swirled in my head.

An elderly looking man with a balding head walked into my room and said gently, "You must be the Charles who has come to join us. I am a Charles, too, Charles Eaglesfield. I am working on the optical communication project and will be your office mate. I am sure we will get along very well." He put me instantly at ease.

"Mr. Eaglesfield, I am glad to meet you. Though I am afraid that I am very new to this research environment. I don't know where to begin," I responded.

"Come, come. Let's start by filling you in on my transmission scheme. John Lytollis is developing a light source. He works for Dr. Alec Reeves, of PCM fame. He is the lead on the Optical Communication Project."

Simply hearing famous names such as Reeves was enough to bolster my enthusiasm for the work. Alex Reeves was the inventor of a technique for converting a continuous electromagnetic waveform to its equivalent train of on/off pulses. The invention of Pulse Code Modulation (PCM) made the digital age possible. In 1936, Alec's invention was the most important step in the practical use of digital communication through the conversion of a continuous wave to on/off pulses. It not only proved the validity of Shannon's theory of sampling rate and its relationship with the maximum frequency component of a continuous wave, but also removed the noise problem through a pure on/off representation. Later, the invention of integrated circuits made the PCM implementation practically and financially feasible. At that time, Dr. Reeves's innovation was too early for useful realization. Transistors had yet to be invented, as well as the integrated circuits that would be needed to make the analog to digital conversion circuits possible in terms of cost and reliability.

While Eaglesfield was working on the reflecting pipe idea, I was collaborating with many individuals. For example, John Lytollis's hollow cathode discharge light source was designed to act as a pinpoint light source of very high brightness. The idea was to create a plasma field in a tubular structure. The hot plasma would generate high-intensity light and then be confined to a small point. In optical communication both the light source and transmission path were the critical missing links.

In the meantime I generated a project to study the design of an overmoded rectangular right-angled bend. The aim was to try to use ray and wave optics together to understand the performance of overmoded wave-guide structures. This became the topic for my doctoral thesis. I was officially registered at the University College of the University of London, studying under the father of the microwave, Professor H. E. M. Barlow, a distinguished pioneer in microwave theory and practice. He contributed to the development of RADAR during the Second World War. Prof. Barlow had been working at a military research station and was instrumental in inventing a weapon for helping Britain to defend itself against the Luftwaffe attacks. As the Battle of Britain entered its critical phase, Britain's fighter pilots were dwindling in alarming numbers. The timely deployment of RADAR helped the defense to win several crucial battles in which the enemy planes often out-numbered the defenders' planes three to one. The radar was able to give early warning to the defenders. Scanning the sky from ground positions, radar picked up the echoes of the incoming attackers, which gave the British pilots the edge needed to plan a winning engagement, despite their numerical disadvantages. Prof. Barlow did not talk much about those days. He only said that the team worked hard and that he was simply in the back room.

I titled my thesis topic "A Quasi-Optical Approach." This turned out to have many hidden difficulties. First, no exact solution is possible. After two years I had completed my experiments and dissertation. I concluded that geometric optics provide a visualization compatible with plane wave propagation, while line or point source emitting spherical or cylindrical waves provide a non-exact estimation of diffraction and leaky waves. This was a transitional learning period that gave me insight into optical waves. The bridging of the limitations of the geometric ray optics in describing optical phenomena, and the complete solution of an electromagnetic description within a defined environment, can be found in only a few cases. No systematic model can be constructed except through intuitive and logically correct approximations. My professor obviously found my struggle sufficiently novel and arguably logical to grant me my degree.

The year 1963 found me involved in the free space propagation experiments in which the HeNe Laser was used to direct a beam to a point some distance away. With a firmly mounted laser, and the beam pointed straight at a distant target, it became obvious why. From a vantage point, we saw a distant light that appeared to flicker. Looking at the laser spot, we realized why the beam danced gracefully around a spot a number of diameters larger than the actual beam size. The changing density of the air diffracted the laser beam traveling through the atmosphere. In the UK, the high humidity made the density of the atmosphere less uniform, hence the beam fluctuated even more noticeably.

The other experiments we performed included those that were repeated in research laboratories around the world, including our confocal lens experiment where we lined up a sequence of convex lenses of equal focal lengths. They were placed at intervals equal to the focal length of the lens. We had to perform the experiment in the dead of night when the air was still. Even then the beam, which was refocused every 100 meters, refused to stay within the aperture of the lenses. Incidentally, Bell Labs decided to go further with the experiment using gas lenses, which was abandoned later due to the difficulty of providing satisfactory insulation while maintaining the profiles of the gas lenses. These experiments could be considered a desperate struggle trying to get light to travel in a controlled manner over long distances. It went ahead because the circular wave-guide project had similar engineering challenges and high costs. The wave-guide systems were to be laid out with only very gentle bends. The gas lens system was not ruled out since the prevailing thought found that the straightness requirement was acceptable.

At STL the focus by that time had shifted to dielectric wave-guides. Dr. Karbowiak wanted to bypass the problem of finding a wave-guide material with very high transparency. He suggested I work on his idea of a thin film wave-guide. I worked with a small team of three engineers to test the feasibility of this idea.

According to our analysis, the thin film should be around one-tenth of the wavelength in thickness when the energy guided by the film should be propagating with 90 percent of the energy above and below the film in free space, and with 10 percent of the energy attached within the dielectric. The first challenge was how to make a thin film. With some experimentation I suggested that we should dissolve a polymer in a solvent and then drop the solution on top of a fluid. By consulting our polymer colleagues in the Lab, we found that polycarbonate had good transparency and could be dissolved in a volatile solvent. The solution would be lighter than water. This meant we could drop the solution on the surface of the water and rely on the surface tension forces to spread the solution, ultimately forming a thin, continuous film. This technique was later known as the Langmuir-Blodgett technique for making ultra thin films and was used in the pharmaceutical industry.

The making of the thin film turned out to be easy and successful. We realized that as the film became sufficiently thin, it had a uniform color depending on its thickness; the color being the interference of light reflected from the top and bottom of the film. As the thickness was thinner than half the wavelength of visible light, the film was colorless. We made various frames to scoop up the film. Eventually we had the machine shop make up a specially shaped frame such that a film will be picked up on an inverted U-shaped sheet. On such a frame, the film laid over the two arms of the solid U. A precut, thin, rectangular slot at the center of the frame was where the film would lay freely and be supported by the sides of the frame. This arrangement allows for the light to be injected into the film through one arm of the U and carried around the bend. The light then emerges from the end of the other arm of the U such that the entry and exit of the light are at the same end but on different arms respectively of the U. For injection and exiting the light beam, we placed a prism on the two ends of the slot with the film and the base of the prism in close contact.

When the film was thick enough, it was easy to see a multiline image of the exit end at the exit prism. This mode pattern was the expected characteristic pattern of a higher-order wave-guide mode. When the film was thin enough, only one mode line existed. Placing the prism in this manner ensured that the light would travel around the bend and not mix with the light injected at the launch end.

One morning we were testing progressively thinner film samples. These films were colorless and we could only gauge their thickness by diluting the solution by a factor of two each time. As we proceeded we were startled because we saw something very odd in the darkroom where the experiment was being carried out. Instead of the light going around the bend, the light seemed to escape around the bend and was illuminating the four walls of the room. The light was escaping as it negotiated the bend. By piercing the film guide around the bend the radiation covered a smaller arc until the entire bend no longer existed. While we did not know how thin our thin film wave-guide was, the radiation around the bend, besides serving as a spectacular display of light, warned us that such a structure would face radiative light energy loss at bends. If adopted in a cable-like structure the ducts must not have sharp bends. A tightly confined wave-guide structure to contain 100 percent of the energy within the material would be highly desirable.

I began a detailed investigation into the loss mechanism of dielectric materials for optical fibers. I was also pushing my colleagues in the laser group to work towards a semiconductor laser with an emission wavelength in the near infrared region, and with emission characteristics to match the diameter of a single-mode fiber. As Dr. Karbowiak had decided to emigrate to Australia to join New South Wales University, I took over as the project leader and my team expanded. We had a small group working on methods for measuring material loss of low-loss transparent materials. A graduate engineer, George, joined me to work on the characteristics of dielectric wave-guides. With his interest in wave-guide theory, he was to concentrate on investigating the tolerance requirements for optical fiber wave-guide. In particular, we needed to define the dimensional tolerance and joint losses. We were proceeding systematically to verify the physical and wave-guide requirements for glass fiber to qualify as a practical wave-guide with the right properties, both mechanically and performance-wise.

Over the next two years, our team worked diligently. We were all novices in the physics and chemistry of materials and in tackling new electromagnetic wave problems. But we made credible progress in gradual steps. We searched through all the available literature, consulted with experts, and collected material samples from various glass and polymer companies. We also worked on various theories, and developed measurement techniques to carry out a host of experiments. Among the types of equipment that we developed, one measured the spectral loss of a very low loss of material, as well as one for scaled simulation experiments for measuring fiber loss due to mechanical imperfections. Special surface wave structures operating at microwave frequencies were used as the means to verify the radiation losses of wave-guides with different types of imperfections. Through the simulation work, George later obtained his Ph.D., based on the radiation characterization work. Gwen was employed at the computer laboratory as a scientific programmer. She programmed the software that gave us the solutions for determining the modes for dielectric wave-guides, as well as for obtaining the radiative characteristics for George. In the now-seminal paper we published in 1966, I should have acknowledged her indispensable contribution to this important task. I was the one to blame for such a gross omission!

The optical loss of transparent material is the aggregation of three loss mechanisms. The absorption loss caused by the material structure itself limits the transparency regions to be wavelength dependent. The absorption loss is due to impurity ions left in the material and scattering loss is due to structural non-uniformity. Since the transparency of glasses for most practical applications was adequate for window glass and ornamental and kitchenware items, no one previously had studied glass transparency and its limits on optical fiber application. I consulted many experts and eventually formed the following conclusions.

1. Removal of all impurities, particularly transition elements such as iron, copper, and manganese to levels in the region of one part per million to one part per billion, is necessary to reduce the loss attributable to impurity absorptions. However, no one has conducted experiments to verify whether the impurity levels will cause the loss to decrease linearly at low concentrations.

2. High-temperature glasses that are frozen rapidly should have a smaller microstructure and more even distribution of nonhomogeneity than low-temperature glasses such as polymers. The scattering loss of the latter class of material is higher.

We were faced with two formidable challenges. The first was the measurement techniques of low-loss samples that are obtainable only in lengths of approximately 20 cm. If the material were made in fiber form, the problem of assuring surface perfection was formidable. The second was the end surface reflection loss. A well-polished surface can have its reflection losses altered by the polishing process. We were facing a measurement impasse that demands the detection of a loss difference between two samples of less than 0.1 percent when the total loss of the entire 20 cm sample is only 0.1 percent. Eventually we were able to construct two-beam measurement equipment that provided reasonable measurement if the reflection loss was repeated each time.

We were lucky to get some pure fused silica (fused quartz) samples that were made through a plasma deposition process. Due to the high temperature involved, the impurity ions were all evaporated during the process; thereby, pure fused quartz was created. The measurement of these samples proved the theory that the removal of impurities lowered the absorption loss. It was noted that impurity analysis in high temperature glass was impossible due to the much higher impurity level involved in wet chemical tests.

In the meantime the microwave simulation experiments were also being completed. The characteristics of the dielectric wave-guide were fully defined in terms of its modes, and its dimensional tolerance limits both for end-to-end mismatch and diameter fluctuation along the fiber lengths. All the results derived from theory and simulated experiments indicated that the theoretical basis was sound. Out of this research came the paper "Dielectric-fibre Surface Waveguides for Optical Frequencies," which was submitted to the Proceedings of the Institute of Electrical Engineers for publication. After the usual review procedure and revision, it appeared in the journal in July 1966. This date is now regarded as the birth of optical fiber communication.

The paper's reviewer turned out to be the then-director of research of the Post Office Research Center at Dollis Hill, London. He was sufficiently interested in the work that he proposed to set up immediately a major program for the development of optical fiber at his research center, as well as awarding a contract to STL focused on taking optical fiber development forward. When the possibilities of optical fiber systems became headline news around the world in the mid 1970s, I received two letters from the public—one alarming and one amusing. The letters were addressed to me, care of the plant in Roanoke.

The alarming one came from someone who threatened me with bodily harm for my devilish and evil inventions let loose on a God-fearing society. This one was passed on to the legal department and the police. Fortunately nothing further came from this letter.

The other was from a farmer in China. I translated it for the benefit of my colleagues. We all had a good laugh.

"I have to work all day in the fields, weeding and raking. My home is a long way on the other side of the fields, too far for me to shout to my wife. It is too tiring to walk back home every time I need her. When I need my lunch, I want to be able to tell my wife immediately that she should bring it to me. Where can I buy one of your new contraptions?"

He had to wait nearly twenty years for his mobile phone.

TESTIFYING IN FRONT OF A HOUSE SUBCOMMITTEE ON DEVELOPMENTS IN FIBER OPTICS

Hitting the Jackpot, Part II

I called a special gathering of my optical fiber communication group and announced, "Our paper has been published in the *Proceedings of IEE*, and we have a letter from the British Post Office inviting us to undertake further work in optical fiber for communication applications. BPO has also told us that a project team will be formed at their Dollis Hill Laboratory."

Our team was tremendously excited. Over the past three years, we had demonstrated that optical fiber for communication was no longer a myth. We had finally supported our conjectures with experimental data that verified our contention. Even if sometimes our experiments were done with sealing wax and strings, we had solid proof to back up our claims. We claimed that by reducing the impurities in glass made with silica, the resulting glass showed an extremely high transparency. So transparent is the glass, that if we filled an ocean with the glass, we would be able to see the bottom of the deepest part of the ocean. We also showed that the fiber should allow light to travel in a single mode pattern such that it could carry on/off messages, just as radio waves could do. Except that light could carry simultaneously many million times more on/off messages than radio waves.

While we were happy with the moral and practical support, we seemed to be the only team pushing this technology. The paper carried a rather obscure title, "Dielectric-fibre Surface Waveguides for Optical Frequencies." It did not appear to make a stir instantly in the communications world, and I was convinced that I needed to take the message contained in the paper directly to the companies that should be interested in this potentially revolutionary approach to communications in order to get them to join forces with us. We needed everyone to be involved to make this project a success. I would emphasize in particular to these companies that optical fiber is made with the most abundant material in the world, namely sand; that sand is environmentally friendly and should be of very low cost. I would tout the special properties of fiber, in particular the lightweight, high-tensile strength, and no light leakage of the fiber. I realized that the ultimate success would depend on the efforts of many people and that I should start recruiting these people as soon as possible. I planned a trip to Asia and the U.S. starting at the end of December 1966.

It was my first return to Asia since I had left Hong Kong in 1953. Apart from visiting my parents and enjoying a well-earned and long-overdue reunion with relatives and friends in Hong Kong, I arranged to visit Japan and Taiwan with a mission of disseminating the importance of optical fiber communication technology for Asia. My trip started with Tohoku University in Sendai, to be followed by Central Research Laboratory of Nippon Telephone Telegraph Corporations, Nippon Sheet Glass Co., and NEC. My first impression of Japan was its cleanliness and punctuality; I found their train system especially impressive.

Gliding along past the neat farms and little clusters of village houses, Japan welcomed me with a bright and sunny day, as the train rolled into Sendai station exactly on time. Presumably they must have known my seat number on the train, for the person was right outside my train carriage when I alighted. He identified himself as an assistant to Professor Zen-Ichi.

Tohoku University was a well-established state university. One of its most notable professors was Yagi, of Yagi antenna fame. Prof. Zen-Ichi was an unassuming elderly professor who was to introduce me to people at NTT. I never learned anything about him except for that he looked after me well. He appeared when I needed his escort, and disappeared when I was occupied with the other people I was to meet.

I was next introduced to Professor Jun-Ichi Nishizawa, one of the most inventive professors in Japan. He is best known for unconventional semiconductor devices. The primary person who had technical discussions with me was a student who was developing a theory of graded index fiber. With a graded index distribution in a parabolic shape, the optical rays would travel at the same speed along the fiber no matter whether the optical beam is launched axially or at any angle to the axial direction. In terms of the performance of this type of fiber, it has a higher information-carrying capacity than a fiber with a uniform core. Since the optical beam is launched into the fiber at all angles, the rays will propagate at different velocities as they travel along the fiber. When on/off pulses are launched, the pulse shape distorts as they propagate along a fiber with a uniform core, while they remain undistorted if the core is shaped with a parabolic profile. This idea was inspired by the con-focal lens concept, and later in the form of the gas lenses. Both ideas were proposed in the early days of optical transmission as a theoretical study. The idea was instrumental in causing the first usable fibers to be made with a graded index core.

The single-mode fiber was my original proposed configuration of the fiber for optical communication purposes. The advantages were clearly stated. However, as fiber fabrication technology advanced, the small core dimension of 3–4 micrometers, in the opinions of manufacturers of connectors, made jointing of the fiber without significant loss of light difficult or even impossible. The first generation fiber with a graded index core of 50-micrometer diameter was adopted as a standard in the early systems.

At NTT, I gave my first talk on optical fiber communications to a room full of researchers. I was most impressed by their profuse taking of notes, but afterwards there were hardly any questions. At this first lecture, it was unnerving to have no questions asked. However, talking afterwards with people from the audience individually, I discovered that they had regarded my talk as frank and informative.

NTT was the citadel of Japan's communications industry. It was then the monopolist supplier of telecommunication services. This situation made NTT's Central Research Laboratory the most important technology center in Japan. At that time they were working on four key areas of research, housed in four multi-storied buildings built as a cluster. Their research achievements inspired and dictated the direction of development in the telecom industries of Japan. The development of the products was left to the competing manufacturing companies. It was a hugely successful practice that generated powerful companies such as Sumitomo Industries and NEC.

During that first trip I visited NEC because it was partially owned by ITT, the parent company of the research laboratory in the UK where I was employed. At NEC I met Dr. Uenohara, a young research project leader about my age. Over the next twenty or more years, our paths continued to cross as we both accumulated valuable experiences and rose in the ranks at our respective companies.

NEC was transformed by the talented and energetic leader, Dr. Kobayashi, who took a war-torn company and transformed it into one of the most successful electronics giants in Japan. Dr. Kobayashi, as the chairman of the Electronic Industries, established an International Prize named the C&C Prize for excellence in Communications and Computers. His personal contribution rivaled the achievements of Sony's Morita, who made his company number-one in the world for audio and video products. Dr. Kobayashi instilled in his people the importance of excellence. My visit to NEC allowed me to appreciate that the importance of group dedication, as demonstrated by their leaders and their people, is why they had the ability to achieve such excellence. Dr. Kobayashi's dedication to his company was contagious and persuasive.

I was in awe of the innovative spirit of Nippon Sheet Glass Company, where I visited next. They primarily made sheet glass for windows, but they took up the challenge of developing the graded index fiber and then branched into making Selfoc lenses with applications for xerography and fax machines. I was impressed at the speed with which the company seized the opportunity to branch out from their traditional products. First, they found a way to make glass rods that have a gradually decreasing refractive index from the center outwards. This was formed by using an ion exchange method. The glass rod could be drawn into a fiber with a graded index core. They also discovered that the same rod could be used to make lenses if it was cut into pieces of a predetermined length, each piece with parallel polished ends. These lenses could be used in the form of a linear array as an image-scanning device useful for copy and fax machines. Over the years the lens arrays created a big scanner business for the company. When I first heard about this at NSG, I marveled at their ingenuity. I remained their friend for over 35 years. The engineer, who was my contact at NSG, became one of the directors of NSG. Now retired, he is co-authoring a book that includes details of the roles NSG played in the development of optical fiber. Their early success in graded index fiber encouraged such fibers to be used in the initial days of optical fiber communication, although the graded index fibers were later fabricated by the MCVD (Modified Chemical Vapor Deposition) technology, a very different process to the one invented at NSG.

Europe was already a familiar stomping ground, even before our historic paper was published. I went to the major glass companies to talk with them about the characteristics of glass and whether they could work with us to increase the transparency of their glasses. We included plastic materials, as well. I had studied electrical engineering at university, and material sciences were hardly mentioned in the electrical engineering curricula. I was destined to reinvigorate my childhood interest in chemistry. My knowledge of both inorganic and organic materials expanded, especially in the properties of materials in a glassy state. I remembered that when I heat, in the colorless flame of a Bunsen burner, a small quantity of common salt (sodium chloride), the flame would turn an intense yellow color. The light absorption of a specific color by glass, due to the presence of iron or copper, was much easier for me to understand than perhaps for other engineers.

Sheffield University in the UK was well known for its study of glass science. I sought their help and they generously provided me with valuable information. They impressed upon me the fact that the glass sciences were not as easy to analyze as crystals. The glassy state could be thought of as a very hard liquid. It softens gradually, floats like a liquid, and can be stretched without internal stress. I learned how colored glasses were made with the addition of a few percent of metal oxides such as copper, manganese, iron, or cobalt. I also learned that glass cooled rapidly would become more homogeneous and therefore would not cause a significant amount of light to be scattered, this being attributable to the small fluctuation of its composition. In particular, they told me that the mechanism of absorption of light due to the presence of impurity elements such as iron and copper were not well known. By reducing the concentration of an impurity such as iron, the absorption of light should be reduced proportionately with the impurity concentration. However, there was no evidence that the absorption would decrease proportionately when the impurity concentration was very low. They felt that to reduce the impurities to one part per million or per billion would be imprudent, assuming that the proportional decrease would be followed. They, for whatever reason, felt this was unlikely.

On the contrary, I felt that the absorption loss should be proportional to the impurity concentration, since at a low concentration there should be little interaction to cause unexpected loss increases. I quote three reasons:

1. There is no fundamental reason in science to say that a purity of 1 part in 1,000 million is not achievable. It has been demonstrated that such purities are not difficult to achieve.
2. There is strong evidence that scattering loss of light through pure glass is negligible.
3. The physical configuration of fiber is achievable.

Glass companies, in general, were not particularly interested in producing very high-transparency glass. They produced special glasses for ornamental art ware and windowpanes. Lead glass has a high density (high refractive index) and is extensively used for art ware, such as in cut-glass vases. Windowpanes are generally made with soda lime and silica. This type of glass is excellent as window glass. A float glass method, invented by Pilkington Glass Company, produces very flat and large sheets of glass of varying thickness. The companies I visited to talk about their potential interest in optical fibers included Pilkington, Saint-Gobain, Schott, Heraeus, American Optical, and Corning. None of them showed any particular interest. They did see a small volume of special glass as a viable product for their markets.

As we delved into an in-depth study of glass, we were entering un-known and unexplored areas of glass sciences immediately. We had no quantitative method of analyzing the impurity content in the glass. We were only able to measure the glass transparency across a limited color spectrum of light. It was a lucky break that one of my superiors in charge of the material area brought back a fused silica sample from Amersil, a company specializing in making silica products. This sample was made by a process called "plasma activated chemical vapor deposition." Ap-parently the high temperature of the plasma was enough to vaporize all the elements, including all impurity ions. While silicon, which has a very high vaporization temperature, reacts with oxygen ions to form silica, it condenses into glassy silica first when the impurity oxides are still in vapor form. In this way, pure silica with little impurity results. We were able to verify from this sample that there were no traces left of iron, cop-per, or of other elements known in the chemical periodic table as transi-tion elements. There remained only a small amount of water trapped in the glass. We used this sample to obtain the transparency over a range of optical spectrum, covering the harmonic absorption peaks of O-H. No absorption of light due to impurities was found.

At one conference I found myself talking with Dr. Robert Maurer of Corning Company. He was the head of research at Corning in charge of the fiber project. I visited him in Corning and met a few young research-ers working on this project. They expressed interest in working with us but did not reveal how they were approaching the subject. At this con-ference, Bob casually asked me about the existence of water in the fiber when different materials were added to the fiber core for various reasons, such as changing the refractive index or altering the melting point of the modified silica glass. Water was a knotty problem since its existence gave rise to light absorption in the spectral region of interest. It is to be noted that water in fiber is still a problem that requires constant attention. This casual conversation revealed that Corning was aware of the water prob-lem and my answer to his inquiry probably indicated to him that I was aware of the water problem, too. This was a typical way of checking the progress of competitors in state-of-the-art research.

We discovered in another early experiment that hydrogen could penetrate glass and create absorption of light. The problem was eliminated by not allowing hydrogen to be trapped around the fiber within the cable. This came about in another casual conversation. One engineer in our company, who was supervising a long-term test of a fiber cable, told me that the transparency of the fiber had deteriorated in a cable under water. After sitting on the bottom of a lake in Scotland for two years, the loss of transparency was negligible, but noticeable. Several months passed before he revealed to me the solution to the mystery. He confirmed there had been an increase in hydrogen pressure inside the aluminum tube that was designed to protect the fibers within the cable. He also confirmed that hydrogen could penetrate into the fiber and react with the oxygen within the fiber to form an O-H bond. It was this that gave rise to the characteristic water absorption. He did not tell me how he carried out the experiment. He was so pleased with solving this mystery that he was glowing with pride, and I was equally proud of the achievement. This is the type of seemingly trivial but significant discovery that excites all researchers.

When the joint work between BPO and Pilkington with STL started in earnest, one of the first tasks was to find the ways and means to make a low-loss fiber. At that time, glass fibers were made by various methods such as pulling glass from a crucible containing molten glass or from a rod with one of its ends heated in a high temperature furnace. In our case we needed to make a fiber with a small core and thick cladding. The interface between the core and cladding could not have any flaws and the refractive index of the core had to be higher than that of the cladding.

We had to decide whether we should start with the high-purity silica rod as a core and find a way to coat the rod with a modified silica layer with a lower refractive index, or use a glass that could be melted in a crucible and make a cladding layer by a second melted glass contained in a concentric outer crucible. The trade-off was difficult. In the high-purity silica case we have no known furnaces that can deliver the high temperature to soften the silica, or to find modified silica with a lower refractive index than silica. In the lower melting temperature glasses we needed to contend with impurities in the glass and from the crucibles. We eventually decided to try the lower-temperature system first. In hindsight, we dismissed the chemical vapor deposition (CVD) technique too readily. It was already successfully being used to provide a passivation layer of silica over semiconductor integrated circuits. When I discussed with the CVD technical expert his method of making silica, he raised doubts about whether such a process could be scaled up for constructing fiber. Little did we realize that Corning had already opted for the CVD method.

By 1970, when Corning announced their experimental single-mode fiber with a loss of 20dB/Km, it had been achieved through the CVD technique. While the first successful fiber was far from being a practical product, the demonstration showed a way to deal with maintaining the purity of the silica in the MCVD (modified CVD—an idea from AT&T research) and OVD (outside VD—the original method of Corning) techniques. We at STL immediately started to look into the CVD technique.

At that stage of development, a personal priority loomed large in our minds. Gwen and I realized that our two children had started to become conscious of their minority status among their peers in school. They often asked us why, with their black hair, dark brown eyes, and flatter facial features, they looked different from their classmates. Gwen, who had experienced being the only Asian in school during her childhood, often talked about our responsibility of letting our children know that they were not a minority in the world.

The dramatic solution came in an unexpected airmail letter. A professor at The Chinese University of Hong Kong wrote to enquire whether I would be interested in applying for a post as the head of a newly established Electronics Department. The post was being advertised in various journals and newspapers. I needed simply to apply and it seemed as if I would get the post. An opportunity arriving at this juncture suited us well. A stay in Hong Kong would at least allow the kids to be part of a community where every child looked just like them. If I could negotiate a leave of absence from STL that would permit me to return to STL during the long summer break, I would not need to be de-linked from the ongoing development of optical fiber communication.

We moved to Hong Kong and I stayed until the first batch of electronics students graduated with their first degree. The employment contract was initially for two years. Though I was anxious to return to research fulltime, I was persuaded to stay on for another two years. I had gained new experiences in teaching and management to add to my career qualifications, as well as being able to pick up my research at STL every summer. We had left the UK late in 1970. In the early summer of 1974, we packed up and left Hong Kong for good, or so we thought. Optical fiber communication had moved to the development stage, and we were emigrating to the United States to continue to work on the advanced development of optical fiber communication systems.

En route to America with the family, we made a stop in Kyoto where a major conference was being held on optical fiber communication. I met some of my old colleagues from STL and we discussed a wide range of issues. I overheard a conversation that mentioned my name, and I pricked up my ears! ". . . Kao, he went off to teach, how did he manage to stay current after just one day here in Japan?" Little did he realize that my excellent connections with STL had remained firmly in place while the rate of progress had been relatively slow during those four years.

Contamination control was too difficult for the crucible-based, low-temperature glass production to handle. The high-temperature silica route was also making slow progress. The problem of joining the two fibers ends with precision and reliability cast doubt on its practicality. Desperate attempts of trying to fill the fiber core with liquid were tried and discarded. The most promising approach was in the modified CVD method in which the core was built up within a silica tube. It was at this stage of the game that I began my new assignment at the Electro-Optical Products Division of ITT in Roanoke, Virginia, a picturesque valley in the Shenandoah Mountains situated in the backwaters of provincial America.

Despite having visited the U.S. many times before on business, after my family landed in Virginia, I slowly started to realize that this beautiful

Roanoke Valley was still living in pre-Civil War days and that Virginia was for Virginians. Women were genteel homebodies, good cooks and housekeepers. Families not entrenched in the town for generations were always considered to be outsiders.

Soon after we arrived, the autumn weather turned the trees of the forest into brilliant blossoms, the leaves looked like flower petals burning from bright yellows to brilliant reds. Against a clear blue sky, the view was unforgettably beautiful.

Many of the locals had never seen an Asian, and we were stared at as we walked by, stopping them in their tracks as they gaped at us, politely turning their heads away when we smiled. It was in Roanoke that my daughter experienced her first racist taunts. "Chinky, Chinky, funny face!" the little boys screamed and screwed up their eyes in an attempt to look Oriental.

My daughter tearfully wanted to know why they were saying such things. Gwen quickly responded, "They are ignorant and uneducated. You need to feel sorry for them that they don't know any better. Their mothers didn't teach them any manners!"

Later, my daughter was refused enrollment in a creative woodworking class because no girls ever were interested. Gwen went and struck a blow for equal rights by confronting the principal.

"If your daughter takes up one of the places in the carpentry class, why, that would deprive a boy from enrolling. Now that won't do. He needs these classes and a girl don't." He laid down this argument and Gwen stuck to her reasoning.

"Because no girl has ever wanted to enroll before now does not mean that none can. This is discrimination!"

Gwen had never said boo to a mouse before living in Hong Kong. She was embarrassed to make scenes in public and never had a ready retort in arguments. Hong Kong had awakened her consciousness. I think it was because, being a colony of Great Britain, its people there had long endured a loss of rights and had learned to show a passive public face to any indignities.

Suddenly confronted with this culture, having grown up in a polite orderly society in the native land of the colonizers, the injustices really began to grate on her. Quite unexpectedly one day she found her voice. She was waiting in a long line for service at a post office counter in Hong Kong. It was hot and stuffy. To her indignation, a large Englishman barged to the front of the line to be served. "Hey," she shouted, "There's a line here. You're supposed to wait in line like the rest of us until it's your turn!"

Startled at the sound of anger, spoken in English with a strong London accent, the man muttered an apology. He turned and left, embarrassed.

"Good for you." said the old lady behind her. "We don't speak English and they just expect to be served first!"

Having found a voice, Gwen began to develop her suppressed assertiveness and has never turned back into a mouse.

So it came to pass that Amanda joined the woodworking class. At the end of the course, of the small rockets each student had to design and make for their final project, hers flew the highest and farthest. I was very proud of her.

Because she had only received a C in mathematics in Hong Kong due to the very stringent grading system there, she was placed in the normal math class at first. The teacher found she had to promote her twice to higher classes so that she wasn't bored. Long before she graduated from high school, the school ran out of math classes for her.

In school, the teasing from a particularly heavyset young man increasingly upset Amanda. He pulled her hair, knocked her books out of her hands, and poked fun at her. I tried to get her to take it more lightly, saying the young man had a crush on her. "Why don't you tease him back and make fun of him, too?"

But she continued to come home in tears. Finally, she punched him. Her right jab caught him directly on his chin—in front of the whole class. Her girlfriends, having witnessed all the past teasing, cheered her on. "Good for you Amanda, he deserved it. He's been asking for it!"

Red-faced, the poor boy sat down in shock and never teased Amanda again. Amanda came home and cried. "I feel so bad. I shouldn't have done that to him in front of everyone. But he just wouldn't stop."

I was surprised that the ITT Roanoke plant was chosen as the center for optical fiber communication development. The only connection to the field at this plant was the name of the plant. It was called the Electro-Optical Products Division (EOPD). Its main product was the night vision goggle, NVG, sometimes known as the Starlite scope, a sophisticated high-tech product for the military. The key component in the NVG was a fiber faceplate that was made by a parallel array of tens of thousands of hollow fibers that act as light amplifiers. EOPD is a division within the Defense Space Group of ITT, working mainly on contracts from the U.S. Department of Defense. Hence I had to get the requisite security clearance at the appropriate level.

ITT's decision to put the fiber development in EOPD was a strategic move. At that time fiber technology was in its infancy. For fiber to meet operational requirements there were numerous challenging hurdles to surmount. In addition, fiber has characteristics that are suited for many military applications. These characteristics are: lightweight for easy deployment on open ground where temporary communication systems need to be set up; high tensile strength, capable of surviving stresses and strains; no radiated radio waves that could be tapped easily by hostile forces. Money from the big purses of the military establishments, the Army, the Navy, and the Air Force poured into the EOPD coffers. This financed, in a very favorable way, the research needed to turn the laboratory curiosities into full-fledged, well-tested, reliable products, meeting stringent military applications, which later facilitated the development of low-cost commercial products. Thus, through a baptism by fire, the real fiber systems were born.

A blue ribbon team was formed, its size and strength growing rapidly into a highly productive force. Jim was recruited from the world-famous Bell Laboratory of AT&T as the project leader and I was given the role as chief scientist. Jim had experience in the design of high-speed transmitters and receivers at Bell Lab and I had experience with fiber technology. Together, we were able to lead the field by developing the first military communication system, which was installed at Fort Monmouth. We delivered a complete system with transmitter, receiver, as well as the fiber cable with our own fibers made through the MCVD process. The cable had the ruggedness and lightweight ability to enable a soldier to carry a kilometer of cable on his back. An equivalent amount of copper cable would require a truck to deploy. We were able to double our contract order input each year from 1974 to 1980, and we collected contracts for projects that made us a jack-of-all-trades in optical fiber, system-related activities. Equipment for splicing two cables into perfect joints involved a considerable amount of trial and error in design: periphery-measuring techniques were practiced. It was not just the fiber cable production that was developed but all of the attendant gear required for copper cables to be re-configured for an optical fiber cable.

During those years of huge profits from the sale of night goggles to the military, of contract overloads and deadlines, my adrenaline flowed fast. I found the work so fascinating, I often could not tear myself away from the plant. Typically, I missed or was late for school performances. I came home late for dinners. My wife reminded me of the running joke about absent fathers—the note on the fridge door to remind the children that the strange man they might see in the morning was their father.

I think I traveled away from home at least one-third of my working hours. There were numerous trips to attend conferences. These, such as ECOC and OFC, were growing over time from small beginnings attended by the original innovators, to big events drawing attendees from an expanding number of professionals in the field. I had been right in 1966 to believe the field needed to attract an expanding number of adherents for optical fiber systems to catch on. It was finally happening.

On days when I needed to make one-day trips into headquarters in New York City, I arose at the crack of dawn to catch the first flight to the city, and the last flight home at night. Longer trips were to Europe, back to visit my old colleagues at STL in the UK, and to other labs on the Continent. My contacts with friends in Japan resulted in many exchange visits.

I traveled so much in fact that Gwen sometimes forgot where I was. She was known on occasion to call my secretary to pass on a request to me to buy something from the grocery store on my way home from work. And the secretary would respond "But, Mrs. Kao, your husband flew to New York this morning; he's not in his office?"

And on the weekends as I was driving the family to the shopping mall, I would insist on stopping at work. "Just ten minutes to check on my results!"

Invariably I would bump into a colleague or two, intent on doing the same thing, and we would have a long conversation that seemingly took only five minutes. The family would be very angry when I eventually returned to my car. One day Gwen decided to teach me a lesson. She simply drove away after waiting for half an hour, did the shopping, and went home, leaving me stranded.

It was a fulfilling time to be working in the center of creative activities. We were capitalizing on our own and other people's research results as the basis for developing the components and equipment for optical fiber communication products. We were working in a military environment where the different stages of development work (R1) to pre-production stage (R4) could be done. We had already done the R2 work and could now move into the R3 and R4 stages.

We encountered many problems related to the fiber performance and its practical use. To illustrate, we discovered that light loss along the fiber increased when the fibers were compressed within the fiber cable jacket. This then led to an investigation of light loss due to a microbending of the fiber, inducing light to leak out of the fiber. After the problem was identified, proper cable designs could be established. We were asked to establish the tensile strength of the fiber to assess whether fiber can be used as a tether for wire-guided missiles. This involved an extensive investigation into the instantaneous and long-term strength of the fiber under stress. We designed testing techniques that allowed us to probe into the problem of fatigue failure (when the stress was sustained over time) for fiber in a humid environment. This was the type of work that allows fiber to be characterized in order for it to be used with assured reliability.

By 1980, we were building a foundation for the realization of practical optical fiber communication systems. Most of our projects were of the

R3 type. In hindsight, we should have been consolidating the optical fiber and fiber cable production, and starting to work with the commercial telecommunications equipment units of ITT. But EOPD was the darling for increasing revenue for the Defense Space Group. EOPD was encouraged to diversify and to continue to expand the range of activities so that it could show more growth in its annual revenue, which was doubling each year; presumably the bigger Group was not meeting its projected revenue target. This was a major error. First, we were trapped into bidding for more and more contracts to meet our order input targets. This move created a shortage of experienced staff that led to contract completion date slippages. Staff morale began to drop.

At the time I thought that the problem was with the local management, but it was also at the Group or even a higher level. The demand for a continuous growth of 10 percent, set by the head office, was too rigidly applied, and was leading to unrealistic forecasts, to a constant swapping of unit CEOs, and to mediocre performance. Looking back, it seemed to be a prime example of the bottom line becoming the only item of importance, blinding men to the bigger picture.

In the meantime, several competitor companies to ITT had made significant investments in focused areas: optical fiber production, components, and equipment for optical fiber communications systems. By 1983, the ITT optical fiber making and military systems remained in Roanoke while the communication system work had been moved to the telecommunication divisions of ITT. By then our competitive advantages were severely weakened due to our over-extended commitment to cover all areas in optical communication. It was this that prevented us from rapidly concentrating on a few chosen areas where we had major strengths. Though ITT was able to adjust and stay reasonably competitive, it was no longer number-one. It was at this point that I was asked by ITT headquarters to become its first executive scientist.

Both of my children had graduated from high school and were away from home at universities—one in Charlottesville at the University of Virginia and the other at Duke University in North Carolina. Gwen, who had spent the years first tutoring high school math for a pittance, followed by two years at Virginia Tech where she had dreams of graduating with a business Master's degree, before abandoning that to run a clothing business for two years, was having her dream house built. She thought we would be in Roanoke forever.

A new phase of our life was about to begin. I went home to break the news to my wife.

WORKING AT A GLASS DRAW TOWER IN THE 1980S

The R&D World

"Dr. Stu Flaschen is on the line for you. Will you take the call?" my personal secretary announced over the intercom.

Dr. Flaschen was the vice president and chief technical director of the ITT Corporation, the person in charge of managing over $1 billion in technical activities for the entire ITT empire. He was a great supporter of the optical communication project.

"What does he want to talk to me about?" This was the first time he had called me directly. "Of course I'll take the call."

Stu started out by saying how appreciative he was of my contribution to the success of the optical fiber project and how indispensable my work had been. I was waiting to hear the word "but." Instead he said, "I would like you to help at headquarters to strengthen the overall R&D at ITT. The board has created a new position, with the title 'executive scientist,' and with a rank equivalent to the CEO of a medium-sized operational unit of around one thousand people. You will be given several years and the necessary funds to do what you think is important for ITT. I hope you will be willing to leave the fiber project in the hands of your highly competent colleagues. In any case, in your new position, you will be able to help them even more effectively."

I could not believe my ears, and I was at a loss for how to react. I had not been prepared for such an invitation, nor was I able to think clearly about the possible scenarios. So, I informally accepted the position on the spot and he asked me to give him an official response by the next morning.

Prior to this phone call, we had decided that we were going to be in Roanoke for a while, and so we had a custom home designed. That evening, when I returned to our new home of only two months, I began missing the house. We were just getting used to enjoying the flowers and bushes around the patio where we often sat for breakfast.

"I'm home. Guess what, I have good news and bad news. The good news is that I was offered a job by Stu and that, after some thought, I decided that I could not turn it down. The bad new is that I want to move to Connecticut where the ITT research center is located or to headquarters in New York."

Gwen had thought that with my recent promotion to vice president at the plant, we would be in Roanoke for at least another five years. We were now destined to continue our gypsy-like existence. Fortunately, it came at a time when both children were studying at universities. The next day I officially accepted the post and became the first executive scientist, with an undefined mission in my head. When I was given a mission as broad as doing something important for ITT, I was essentially being given a blank check and the freedom to do anything, so long as the outcome benefited ITT. The major challenge was to paint a rational picture of the future forces that would propel science and technology forward. I had to first understand the obstacles that were deterring the progress of science and technology, and to estimate whether the progressive removal of the limits set by such obstacles could be done.

Hanging in my bedroom is a large photo of me, taken on a green lawn with lots of trees, with me reclining against a tree in a relaxed posture. With pencil and pad in hand, I look as if I am deep in thought. This was the PR picture that launched the news of my appointment. It said in one caption, "We give him money and time to create a better future for the world." At the beginning of this assignment, I looked at this portrait often, not because I was vain, or because the picture was asking, "Are you dreaming?" It was sending me a cautionary message; that I should take realistic steps and proceed without letting the legacy of events cloud my judgment. To help ITT's overall technological progress, I needed to address the basic issues of R&D on the management of technology, setting innovation boundaries, and assessing the impact of technology on new business developments.

As Stu's proposal of my appointment, duly approved by the chairman and CEO of ITT, reached the personnel director, I heard through the grapevine that the personnel director did not agree with my being promoted to a line management position, but that he would approve my appointment to headquarters staff, with the equivalent rank of a CEO of a mid-sized business unit within ITT. He was reported as saying, "Charles is not managerial material." Was that an excuse to impose a glass ceiling? Regardless, I was pleased with the eventual solution, as my pay was equivalent to an appropriate CEO grade. I should thank all the people concerned, because the executive scientist post allowed me to push forward my aspirations, and later to open the way towards my second career as the president of a university; probably one of the most demanding management posts in any organization.

My temporary abode, when I started my new post, was a rented apartment near the Advanced Technology Center (ATC), the central research unit of ITT located in Connecticut. I was to move there ahead of Gwen, since we had to look for a suitable house to buy or to build. We were determined to have a custom-designed place. After all, we could not move from our dream house into just any ordinary house. In the meantime, we bought a lot in Trumbull, a small town not far from Yale University. We found a builder who would design and build a house to our specifications. It was to be a modern California-style two-storied house. It was built on a well-wooded lot with a gentle slope that became steeper down towards the back. A sizable wooden deck at the back of the house allowed us to sit outside and relax comfortably, and Gwen could watch me laboring with our motorized mower. When we returned after a trip to Germany, we embellished the platform by installing an awning that we had bought in Germany. We actually bought two such awnings, one of which was never used in Connecticut, but was later installed to give us shade over another deck at a New Jersey home, which we bought when I left ITT.

During the first year of my tenure as executive scientist of ITT, I recruited two brilliant young men and formed a team of three. One of them was a fresh Ph.D. from Caltech and the other was someone with outstanding analytical skills. The program I initiated was aimed at answering the question, "What is the highest on/off switching speed that a semiconductor device can reach?" The answer would allow research projects, involving the use of semiconductor devices in communications equipment and systems, to be assessed for their worthiness for support. Effectively we have a means to rank R&D projects in order of their competitive merits against possible future products of other manufacturers. It would be an important management tool for the allocation of R&D resources.

The fundamental speed limit of any semiconductor device is the limiting speed of the movement of an electron within a semiconductor. Thus, the problem can be reduced to that of finding the movement of an electron in a bulk semiconductor. The solution is to be found through the application of Schrödinger's wave equation to the problem. After two years the team presented the results showing how the movement of electrons proceeded within the crystalline atomic structures of a typical semiconductor, such as GaAs. It showed that electrons would be scattered as they moved forward under an electric field. It was concluded that typically the free electrons arriving at their destination are a small fraction of the total electrons moving through the material. Those electrons scattered by the material structure were trapped in forbidden energy bands and later would leak out. As a result, the pulse of electrons would be broadened instead of remaining as a tight bunch. Further theoretical analysis showed that the limiting size of a pulse propagating through the semiconductor material is unlikely to be shorter than 10E-11 seconds. Thus, for nonquantum semiconductor devices, the pulse repetition rate is limited to below 1 terabit/sec.

At the same time, I set up a team of people from ten universities and research units. I thought that if I could fund a leading expert in different, but related areas to the electron motion problem, I would be able to create a stimulating environment for the researchers to mutually influence each others' thinking, as well as to save lots of money by distributing the work to those experts who already were spending money in those areas. We were funding each team at $200,000; this was equivalent to paying the loaded cost of a qualified engineer at ITT for one year. I estimated that we were getting substantially greater monetary worth for the valuable new information that resulted than the actual cash payments. For example, the team headed by two world-famous professors at MIT were jointly involved in our project; they had equipment in their lab that allowed femtosecond speed electronic events to be analyzed. In our project this equipment was available for experimental verification of our theoretical studies. Since we were dealing with the motion of electrons occurring at high speeds over a short time interval, their interest in this area allowed us to tap their expertise.

We had a couple of conferences attended by the team leaders and one or more team members from each team. The fact that all of them came to these events without persuasion was a good indication of the value the participants gained at such gatherings. The information that we gathered and the knowledge that we generated significantly helped me to see how the field of electronics was developing. I can confirm now, many years later, that the way we moved forward with this project was the correct formula. We also realized the benefit of an effective close coupling between industry and university.

By 1986, at the end of my three years as executive scientist, I returned to the main research laboratory of ITT in Connecticut. At that time, ITT was undergoing rapid changes. Alcatel, a French company, was in negotiations to purchase ITT. This would instantly make Alcatel one of the largest telecommunications equipment suppliers in the world, second only to AT&T. With the presence of ITT units in almost every country in the world supplying equipment to the telecom industries, Alcatel would automatically become a multinational company overnight. When the agreement was signed, I was invited to consider heading the R&D laboratory in the U.S.

The interesting tools that I was instrumental in creating during my tenure as executive scientist did get into the R&D management system of ITT. The technical director and his planners formed a number of task forces to map out the performance trend lines of key components that controlled the overall performance of the system products. This effort helped ITT to introduce highly competitive products into the market in a timely manner. Such a forecasting effort enabled the product planners to define the specifications of the products with a greater assurance of success in the market. Nonetheless, the twists and turns of mergers and acquisitions did not allow the work that I did to yield its full benefits for ITT. However, if I had not had this experience as executive scientist, I could not have followed my subsequent careers as competently or as well.

My project on "How fast can a semiconductor device operate?" defined the upper limit of speed that a system could achieve. It established the limiting performance of all systems involving semiconductor devices. It effectively defined the high price to pay for reaching this limit. The knowledge gained from this project was used as a selection tool for sorting, from amongst the numerous proposed systems and components projects being proposed, into three categories. Those that achieved too low of a performance or too high of a cost were eliminated. Those that achieved too high of a performance were carefully handled because the cost potentially could be very high. The products that allowed competitive performance to be realized, without stretching the technology limits and cost target, ended up in the preferred group.

My other project was written up as a book, which attempted to explain why high tech-business must be handled differently. It was produced in draft form while I was in Germany when most of my colleagues in the Research Center at ITT's Standard Electric Lorentz (SEL) were on their long six-week summer vacations. I had no interruptions from meetings or phone calls. I wrote in the preface of the book eventually titled *A Choice Fulfilled*, "My one time colleague in British industry, turned academic professor, and now rector of a famous university coaxed me at all his lectures on highly technical topics, to believe every word that he uttered. Yet every time I tried to convince my audiences that technology had changed our business world, specifically high tech business, no one appeared to be convinced.

Several years ago when I was working in Germany, I took a deep breath one day and wrote down the gist of my arguments. I decided that if I expand the salient points, I should be able to demonstrate unequivocally my convictions. During that hot summer, when most of my German colleagues were on vacation, I roughed out my first draft."

That was in 1983. When the book was eventually published many years later in 1991, I added a couple of chapters to my original draft and the text was considerably improved by a helpful professor in the Business School. Gwen also embellished each chapter with a thoughtful cartoon. Quoting the preface:

> A look at the titles of the chapters might convince the readers that much of the book is about technology and not too much about business. This is true. However, the first chapter, "Opportunity Unlimited" reveals the exciting vista of the new brave world in which technology has wreaked havoc and has turned the high tech business world seemingly upside down. Chapter nine brings us into a tantalizingly almost real participation in a conference on "Technology Transfer." Questions might spring to mind which the reader jumps to his feet in frustration to ask and may well find answered on the next pages. The rest of the chapters are individual examples to illustrate the causes and effects of technology on business. The transistor miracle, the computer revolution, the optical magic and the information age are all there. The multiple impacts of technology are highly revealing; a new business world with an over-abundance of opportunities, ready for any choice to be fulfilled, is awaiting.

To illustrate more clearly some of the arguments on why business is impacted by technology, I quote, first, a few paragraphs at the start of chapter one entitled "Opportunity Unlimited," as an illustration.

> Br . . ., Br . . ., the telephone at the bedside rings. The guest in the plush room of an international business class hotel picks up the

phone. He hears the computer voice saying mechanically: "This is your 7 o'clock wake-up call. A very good morning to you. The outside temperature . . ." He replaces the phone and switches on his TV and tosses and turns a bit to wake himself. At 7:30 he places a phone call to his secretary to give her several messages which he could not send through the electronic mail system from his portable personal computer. "It is such a nuisance that not all hotels have the electronic mail connections," he says passionately. "Next time don't book me into just any hotel," he adds indignantly.

This scene illustrates clearly several customer needs and market opportunities fulfilled and not fulfilled. It is just a tip of an iceberg representing vast and real, but hazardous, market opportunities reachable through the use of appropriate technology. Hazardous because this customer may not be representative of a large number of customers with similar needs and yet his demand is influential. Should the hotel invest in an electronic mail system so that it can retain this customer? Was the wake-up call service appreciated? An automatic switch to turn on the TV may be quite adequate. Should video contacts replace the need to travel and, hence, displace the need for business hotels altogether? How should a TV manufacturer customize the sets for hotels and thereby secure that particular market niche? What should computer makers and software writers do to cash in on the wake-up call market or perhaps to ensure that they can gain a foothold in the electronic mail service market?

Our starting point is, therefore, a small but typical case of opportunities brought about by our real needs and our ability to meet them through the use of technology. No wonder decision-making is becoming infinitely complex for all users, service providers and manufacturers alike. How can a new product be envisaged and what does it take to introduce it successfully in the marketplace?

What infrastructure is needed to support this new product and how can the users be encouraged to take full advantage of it? The answers to these questions imply the necessary conditions for successful exploitation of any new business opportunity.

The solutions to these questions are, in general, totally elusive, but for specific cases, they can have unlimited opportunities. This situation transforms the very basis of established business practices and calls for new management techniques in all aspects of operations, engineering, marketing and sales. It is no wonder that success in personal computers and electronic gadgets for offices has been so sporadic and unpredictable. We shall begin by looking at what changes technology has wrought and why we have a confluence of achievements which should encourage the emergence of a whole host of good business opportunities.

Perhaps the most significant factor that technology has brought to us is the means for products to be tailored to meet and satisfy individual needs at affordable prices. This is a bold and sweeping statement which must be carefully explained and substantiated.

I find the opening words that I wrote over fifteen years ago uncannily precise and to the point. Looking at the actual developments of the market that had already taken place, my words could be interpreted as correctly capturing the essentials of how technology has impacted the market. The computer is now an ever-improving product that can deliver the individual user's needs from this grossly overdesigned machine to encompass all users needs, and it still can be made at a cost affordable to every user, even if the user only wants to use a single feature out of many, many available features. This type of tailoring of the machine for the individual is one way to meet the marketing demands for an affordable machine tailored to the specific needs of a customer. The business model is indeed impacted by technology.

The Internet revolution was still in its infancy when we entered the twenty-first century. We can see clearly that the advances in electronics and photonics lowered the cost of storing, processing, and transmitting information, to the extent that new services meeting our communication needs may be provided at an affordable cost and improving our ability to create valuable additions for our efforts. We coined the terms "information age," or "knowledge-based economy," because we can foresee a big change coming. The opportunities of taking a gigantic step in raising our productivity, and in achieving a much larger global economy are at hand. This potential economic growth is attributable to the new opportunities made possible by technology. As I mentioned in the book, the impact is as massive as the invention of the printing press.

On the other hand, the world is not an egalitarian state. The Third World, as we refer to the less-developed areas, still faces steep odds in reaching a decent standard of living. The Third World remains in the same state even now after I wrote the following script fifteen years ago.

> Within the global village, many well-guarded oases have been created and occupied by the lucky and the enterprising. Outside of these are the hardier areas. Where habitation is possible, swarms of the less fortunate and the less privileged scrape together a mere existence. Some parts are still fertile land which were undoubtedly oases in their times. Overpopulation and poor organization have created a situation of over demand for the products from the land. War and natural disasters have not helped. These are areas struggling to arrest the decline against overwhelming odds. These are the areas trying to play a catch-up game with scarce effective means. These are the Third World.

> To the people of the Third World the consequence of abundance created by technology is that of apparent hopelessness. The prognosis is extremely poor.

Of course, the global situation at the dawn of the twenty-first century has produced some changes. Three new technologies, developed through years of research, have reached the stage of maturity for a wide range of applications. These are information technology, biotechnology, and nano-technology. I wrote a paper on the outlook of these three technologies at the beginning of the new millennium. The following is the main body of this paper.

The Outlook of Information Technology

Information technology is the means to deal with the transport, storage, and processing of information so that the needed information can be accessed at anytime, at any place and by anyone. Information technology allows tools to be created to help us access and deliver the needed information instantaneously to any physical location wherever we happen to be. So far we have two important means to distribute the information globally and locally. These are by the wired and wireless transmission network. With the invention of optical fiber, the wired transmission networks are being built with very large bandwidth that can meet most of the normal needs. The wireless transmission takes place in the free space and is designed for providing anywhere access, together with the wired transmission network. While the local bandwidth available for wireless transmission is more restricted, efforts are being made to extend the available bandwidth.

The development in Information Technology is progressing rapidly and successfully. We now have a global network that reaches all the major cities of the world providing the desired inter-connectivity for billions of users for both voice and data. Central to this success is the existence of the Internet and its World Wide Web for data and the robust telephone system for voice communication. The merging of the two is in progress. Our effort to transport and store information at anytime, at anyplace and by anyone has reached its first milestone. We achieved the interconnectivity on a global basis for voice, vision and data.

We are now driving towards the second milestone aiming at making the global network usable for applications that will add value to our life. Our first challenge is to make the network secure, reliable and capable of delivering services with quality assurances. Our telephone system has set a very high standard in the provision of a reliable service. All of us are used to expecting the telephone service to meet 99.99% availability or better. However, for data services we are in the early days of understanding of our exact needs. Currently, we access the data services on the Internet. We connect to the Internet via an Internet service provider, ISP, via telephone lines that has limited bandwidth. We try to reach services offered by service vendors without the assurances that access has the bandwidth to accommodate the demands. We have a lot to learn before the delivery of services would meet the requirements needed to satisfy all the customers. For example, a trading company could deal with its access shortages by having many telephone lines to cope with the large number of simultaneous calls expected during peak trading hours. This same trading company would need to have sufficient bandwidth capacity to cater for the data flow. Customers obviously would find certain types of delays unacceptable, and be concerned with security issues relating to transaction and payment. A solution to such issues is difficult to formulate, since customer expectations have yet to be formed. Yet the trust between the customer and the service providers must first be created.

It is argued that improving information flow is fundamental to economic growth. If we have more pertinent information at the right time we can make better decisions for achieving a higher value addition. It also implies that communication to and from suppliers and purchasers must be well synchronized.

At the beginning of our quest to use Internet for improving information flow, our network and its infrastructure are far from complete. It was the glamour and over-expectations of the investment bankers that led to a flurry of activities in the formation of e-commerce that then eventually led to the bust of the dot.com bubble.

The availability of venture capital from the investment banks allowed a large number of entrepreneurs to establish high-risk ventures in e-commerce. Expectations in such businesses of super fast growth and return, led to market frenzy. Reliable business models were thrown out of the windows in favor of untested models based on the assumption that fast information flow increases the rate of value addition with long term benefits. This led to the testing of an assortment of e-commerce from selling raw information, providing portal information, consolidating niche information, B to C, and B to B supply systems, advertising, auction and etc. The quick footed equipment makers, network builders, ISP and data center builders saw very fast growths. The size of companies established ranged from super large ones to one-man bands. It was a surreal time during which only a few emerged as multimillionaires and many failed. It left an experience base, both wrenching and hopeful, for the information industry to rebuild itself.

Nonetheless, Internet allows multi access interactive, gathering, sorting, and processing of information amongst unlimited number of participants. A new epoch in which our individual and/or group effort, augmented through fast interaction and coordination, should yield an increased value. The contribution to wealth creation, relying on sensible, well-tested business models that are based on knowing the speed and trend of social changes, is significant.

Information technology has by and large solved the bandwidth shortage and the storage and processing capacity problems. With the continuing advances in semiconductor technology and display techniques, the challenge ahead is to deal with intelligence level of storage and retrieval issues. Our information and knowledge are stored without the context of the how's and why's, specific information and knowledge is generated or rooted. At the same time, when we ask a question we seldom mention our intent. Moreover, our language is evolved with ambiguity and designed to be inexact. How should the information be stored and how should the questions be addressed are important issues that remain to be tackled. Otherwise, we are unlikely to be able to get the information we are seeking. In fact the solution can only come from many sources between many people, just as we solve it now, namely, discuss, consult, speculate with people with special areas of interest and knowledge.

It should be obvious that information technology will lead us to another era of our progress. While the way ahead is clearly seen as an evolutionary process, the restructure process can be traumatic. For example, teaching and learning of an ever-increasing and changing volume of knowledge may require a different way of educating the young. The students need to learn how to learn and the teacher to learn how to teach and the parents to learn how they learn their role in the process. The work force needs to update themselves with continuing life long education. The retirees need to have a place in the community especially since they are most likely to be more physically and mentally sound than in the past decades. The methodology of teaching and learning will evolve. We are going to be more interdependent and have more levels of intermediaries to provide us with assistance.

The Outlook of Biotechnology

For the first time in human existence, we are facing the issue of being capable of delving into bio-molecules. The chemistry of biology is known commonly as biochemistry. It is the study of the chemistry of organic molecules. The name biotechnology is created to cover the area of modifying or even making bio-molecules, including the living molecules that can self multiply under the control of the DNA.

Ever since the discovery by Watson and Crick of the double helix structure of the DNA molecule, the door is open for the creation of living molecules. Recently, the entire human genome has been deciphered. This gigantic task resulted in our having a map that delineates the various command functions that guide the formation of a human being. The way the DNA functions, is still far from being understood, but we are already able to link the role of certain sequences in the DNA molecule to disease and human growth. For example: certain genetic disease is due to a fault in a specific gene sequence. By correcting the defect in the sequence through gene manipulation the disease might be cured. Through research, certain gene sequences have been identified as controllers of body functions such as aging, or fatness, etc.

In the botanical world, the manipulation of the plant characteristics has been done for a long time. The method of cross-pollination enabled new fruit and flowers to be created. A famous story exists in Holland describing the struggle to grow a black tulip. A more practical and down to earth project undertaken was the development of a new strain of rice called the miracle rice which was genetically modified in order to accelerate the growth of the rice plant, so that such plant could be planted in colder regions and still be harvested twice a year. This project is aimed at increasing the supply of rice. Genetically modified foods are now routinely grown and eaten. The controversy on GM foods arises over the incorporation of insecticide into the genetic makeup of the plant, or the incorporation of toxic ingredients to slow down the ripening of the fruit. The verdict is still open whether such genetically modified food could produce long-term undesirable effects to humans through the long food chain.

Cloning has been successful of mice, sheep, and other animals. Identical replica has been successfully produced. However, cloning a human being has been criticized as religiously unacceptable and is banned so far. Such a controversial experiment is likely to be carried out soon. Even more controversial is the production of human spare parts using stem cells gathered from fetal tissues. Apparently, stem cells can create almost any organ. Thus a human ear has been seen grown on the back of a mouse.

The day for the engineering creation of a human being is a long way away, but the way to elongate the life of a person through genetic modification, and through spare part replacement of any diseased parts such as kidney, liver, etc., is already possible. "Should such experiments be conducted?" is a political and a religious question. Let us take a practical question of a premature baby born after 15 weeks when the baby's brain has yet to develop and when the baby's lung is full of fluid. Should the doctor abandon the life of this baby? The doctor has to make the decision. Such situations are forcing us to re-visit the traditional view of life.

Biotechnology is the technology that brings the issues of morality, religion, equality, human rights, and environmental protection into the forefront. It forces us to confront these issues squarely. Biotechnology can solve some of the problems, but the social norm today may not allow the necessary changes to take place in time. A significant paradigm shift in societal values is needed.

The Outlook of Nano-technology

Nearly 30 years ago I saw for the first time a small model of a rotor built on a piece of single crystal silicon. Various lithographical technologies were used to make this 3-dimensional object. The researcher told me that silicon was one of the best materials, because of its physical properties, for fabricating into precision mechanical parts. The entire structure could not be seen except through a high power microscope. The first application was in using silicon groove as a channel for lining up two optical fibers. The crystalline silicon offered the possibility of achieving dimensional accuracy in the order of nanometers.

This was an early example of nano-technology where the material parts are measured in nanometers. One nanometer is approximately six linear lengths of hydrogen atoms. At these dimensions the material properties differs from their bulk properties. We now use the term nano-technology to mean the technology that enables small mechanical devices with dimensions around 10's of nanometers or less to be made. Composite material using the nano-tubes could have important physical and chemical characteristics and could achieve very high strength to weight ratio. Another example is the many sided carbon crystal ball. It has many unique properties.

We are at an early stage of finding usages for nano-technology. At the present stage of development we are experimenting in the formation of nano-scaled materials. As mentioned earlier, we can manipulate atoms one at a time. But this does not mean that we can make a millimeter long line of atoms, since we have to move millions of atoms to achieve this feat. A cubic centimeter of a new material built one atom a time is just a dream, unless we can do this process at very high speed and in a controllable way. The importance of being able to self assemble in a controllable way may allow useful material to be built. So far the only successful growth of a layer of material at one atom thickness, in a self-assembly manner, has been achieved by epitaxial growth of material on a single crystal substrate. This technique is known as atomic epitaxy. The principle is to allow surface reaction that allows only one atom to be absorbed at a planned electron-coupling site. Once an atom is absorbed at that site, no other absorption can take place. This ensured that only one atomic layer of new material is deposited.

Other single atom growth techniques are found in single crystal growth. Carbon tubes and carbon balls with different 3D configurations have been successfully grown. In general, the materials so produced have been shown to have unexpected mechanical and chemical properties. In the biotechnology area, the crystals of proteins have been demonstrated. This means that we may be able to make new proteins.

So far nano-devices, such as a motor, a magnetic sensor, have been demonstrated. New crystals allow new material properties to be studied. The reactions of such materials with the environment may allow new sensors to be built. Since the dimension is so small many measurements instruments can be made that can be inserted into places such as in the arteries to monitor blood flow rate and other vital signs for health diagnosis purposes.

We may speculate that nano-technology could lead us to making materials that have mechanical properties far superior to existing materials. If the material is more durable and has a much higher strength to weight ratio than titanium and can be made at a low cost, we shall see the designs of everything that enriches our life dramatically improved. If the materials are non-polluting and durable we may even be able to solve some of our environmental dilemmas.

Postscript

The developments of three new technologies, Information Technology, Biotechnology, and Nano-technology, are in their infancy. Because they are new to us, we are wary about how these technologies should be applied, especially since the new applications could change our social norm. We could face a paradigm shift in which our long held beliefs are being challenged. The rapid information flow can destabilize the community when the lack of understanding between the rich and poor, the educated and the illiterates, the supplier and buyer can be the cause of major conflicts. The whole economic stability could rapidly collapse and mass destruction might be unavoidable. The bionic human being is nearing a reality. If everyone could live to double the current expected longevity and if the genetically modified human beings, animals, and plant life turn out to be not as expected, who will be able to pick up the pieces and restore law and order? It is highly likely that the laws of tomorrow are unlikely to be enacted fast enough to bring back stability. The new man-made materials are infinitely usable. With their presence, human ingenuity can be exercised with a broadened scope. All that is impossible could well become possible. The question is: where are these technologies leading us?

Maybe one day, the earth will be hit by a meteor and rendered uninhabitable by human beings. Maybe one day, the sun will die and be transformed into a black hole. Who would then care? But the human spirit may exist in an ephemeral state and return in its next incarnation. I expect that we shall meet again to recount our experiences.

Having told my story of how the events developed and how research on an aspect of science proceeds, it should be clear to the reader that it takes many years for the germ of an idea to become a real product. It takes many hands to move forward and many minds to add more knowledge to the pool of human wisdom. It takes persistence and a shared belief.

That I made the first tentative steps as a pioneer does not entitle me to royalties from the manufacturer of the final product. Many in the public have asked me this question and commented, "Wow, you must be rich! All those royalties!"

In any job, the employer is entitled to own the time and labor of the employee, for which a salary is exchanged. In research and development jobs, it is no different. The company owns the ideas conceived in their research laboratory. If the ideas are useful to the company, I might be awarded a bonus at the end of the year, or a handsome pay raise or promotion. One could be cynical and say that the rewards are there to ensure that I do not resign and take my ideas to a rival company. Not everyone is lucky enough to have a job that they have an intense interest in, so that work seems more like play.

It appears that the tools for R&D management, on setting innovation boundaries and for accessing technology's impact on businesses, are even more relevant in the twenty-first century. I look forward to seeing how these tools are adapted to the new environment. Already we can see the importance of working globally and in an interdependent manner.

My description of the three emerging technologies is a precise preview of what further R&D might need to be carried out. It is a speculative outline of a research and development agenda for the future. The account of my experiences merely serves as a statement of the unknown confronting us, and challenges us to provide the needed solutions for the future.

VICE CHANCELLOR CHARLES KAO OVERLOOKING THE CUHK CAMPUS

Vice Chancellor of CUHK

Why the head of a university is known by different honorific titles is steeped in historic and legal legacies in various countries. The title of vice chancellor is broadly used in England and, hence, has been used in almost every university throughout the British Commonwealth, while president, rector, chancellor, and director are other well-recognized alternatives.

During my tenure on the teaching staff at The Chinese University of Hong Kong (CUHK), I had the opportunity in 1972 to attend the Commonwealth Universities Congress in Edinburgh. The congress, held every four years at varying locations, is for the heads of universities to exchange ideas. Staff members from various universities contribute invited papers. The subject of my paper was "On the Ecological Impact of Land Reclamation in Hong Kong." It was the first large-scale academic "boondoggle" that I had the pleasure of attending. Not only was free liquor flowing generously, but the more junior staff members of the university systems could rub shoulders with the often invisible leaders of the institutions.

One day, when I was chatting with one of the newly appointed heads of a Canadian university, I decided to ask a rather probing question. "What do you intend to accomplish for your university as a newcomer?" He stopped momentarily and answered in a pensive way: "The Canadian universities are all well established. As a new VC, I intend to concentrate on making one change. A change that I feel would improve this university. I believe that I am entrusted to accomplish such a task."

He continued, "A great university is built through the combined efforts of the staff and the students. I am the shepherd who must provide the space for everyone to grow, while guiding them, in unison, to reach for an additional dimension of excellence."

I was duly impressed.

Presidents, rectors, and vice chancellors were everywhere. Having shed their aura of office, and without their usual trappings at their official workplace, they all looked and behaved like regular folks. My participation at this event, without my purposefully seeking it, undoubtedly helped me later take on such an office and its responsibilities. At that time, a long-term academic career was far from my mind. This was meant to be a short break from my industrial self.

One day in the spring of 1986, a letter marked "Personal and Confidential—Not to be opened except by the addressee" arrived on my desk. My secretary had laid it on my desk in such a way that she must have suspected something but, being a professional, she made no comment other than placing the letter so that I would not miss it.

The letter informed me that the vice chancellor of The Chinese University of Hong Kong would be retiring in a year's time and that the chairman of the selection committee was inviting me to apply for the post.

The letter was a surprise, even though just one year before I had received an honorary degree of Doctor of Engineering, Honoris Causa from the same university for my contributions to the engineering industry and for my contributions to the establishment of the Department of Electronics at CUHK fifteen years earlier.

The honorary degree ceremony was an extremely pleasant affair. Three other honorary graduates and I received awards. We were lodged in one of the top Hong Kong hotels in Central, in a suite of rooms. The rooms had a magnificent view of the harbor and I could appreciate, from that lofty vantage point, all of the many changes that had occurred over the intervening years since my last sojourn in the city.

The solemn and dignified ceremony began with us marching into the auditorium following a long column of teaching staff in their academic finery of varying designs and colors; a clear demonstration of the strength and diversity of talent trained all over the world. Those familiar with the robes could readily tell the pedigree of those teaching staff. The honorary graduates marched just in front of the vice chancellor and the chancellor, who by tradition was the Governor of Hong Kong. I had marched in such processions during my earlier four years at the university. As a lowly teaching staff member I had found it a hot and tiring affair to sit and watch every graduating student parade before the dais. Now that I was sitting up on the stage to wait my turn to bow before the chancellor, the pomp and circumstance of the ceremony impressed me with its solemnity and dignity. It conveyed to all the graduates a sense of achievement and a feeling of appreciation for what the teaching staff had done for them. As an honorary graduate, I appreciated the honor the university was bestowing upon me. I was given the additional honor of speaking on behalf of the other honorary graduates.

At the dinner given by the university in recognition of the honorary graduates, there were more speeches. The vice chancellor gave no hint that he and a number of senior staff were looking for a successor. Whether I was already a candidate in their minds at that occasion I would never be able to verify.

Immediately after I received the unexpected letter, it was less of a surprise when several emissaries sent by the university came to call. Their purpose was to familiarize me with the many aspects of how the university was progressing, and to urge me to apply. "The University honored you last year. There is no such thing as a free lunch, you know! We expect something in return."

They assured me that I would be one of an unknown number of candidates, and that no one would know that I was putting my name forward as a candidate. Thus assured, I took counsel with Gwen. We had already discussed the matter many times since receiving the letter. Though I had often examined all the many possible directions that I could follow in my career, I had never given much thought to an academic path, let alone to being the head of a university.

We took the simplistic approach as I finally dropped my application form in the post. We argued that the invitation to apply for the position was not the same as agreeing to take the job. There were other candidates to be considered and we should take it one step at a time. In any case, I had been given a position at ATC, the research laboratory of ITT in the

United States, after my tenure as executive scientist ended. I was already busily engaged in organizing the advanced research areas.

I was in Europe when I received a phone call from Hong Kong. The secretary of the Chinese University was on the line to pass on the message that the dates for my interview had been set up and that I should be prepared to fly to Hong Kong and stay for three days. Could I be present in Hong Kong for the interview? I checked my diary and found that I could fly directly from Europe to Hong Kong, since I would again be scheduled to be in Europe.

There was no anticipation on my part about a selection committee interviewing me, or that it would end with an immediate request for me to decide whether I would accept their offer. I was quite nervous when I entered the interview room. Invited by the chairman, a prominent banker I later discovered, to sit at the only vacant chair in the room, I faced a panel of unfamiliar faces. The chair was placed next to the chairman's seat while the other members sat in a row opposite the chairman and me. They tried to put me at ease as the council members directed a number of questions at me in turn. I answered their questions in English, Mandarin, and Cantonese, according to which of the languages the questioner used. Perhaps they were testing whether I could deal with questions in my poor Cantonese. I remembered that my mouth felt unusually dry during the interview, and I suppose I was talking too fast, concerned that I wouldn't have sufficient time to make my case clearly. I had not needed to explain myself before a selection board for decades and I felt nervous and a little panic-stricken. The roles had been reversed for so long that I was more used to asking and interrogating, and I had forgotten what it was like to be on the receiving end of the questioning. I was relieved when the session ended with warm handshakes all round.

Driven back to the hotel in a comfortable chauffeur-driven car, I was happy I could relax before they would call to let me know the program for the next day. My jet lag caught up with me and so I took a nap. At about 9 p.m. that night, I received a call from the chairman of the board.

He said, "At the meeting of the council of the university, the council members unanimously voted to offer you the position of vice chancellor beginning in the next academic year. Tomorrow you will meet a delegation representing the staff of the university and another delegation representing the student union. These are important meetings. The purpose of them is for you to have a dialogue with these two groups of interested parties before the appointment of the position of vice chancellor. Please think through whether you have any further questions that you might like to have answered. Please let me know whether you will accept our offer by tomorrow morning."

He added as an afterthought, "I understand that you already were given all the materials regarding the responsibilities and the terms of service."

The chairman was very brusque and formal, and I was somewhat taken aback. Later, when I got to know him better, he turned out to be kind, considerate, and a decidedly no-nonsense leader. He helped me considerably with advice and suggestions for tackling some of the most difficult situations during my nine years as vice chancellor. He was, by nature, a taciturn character of few wasted words. Ladies who were seated next to him at the many formal dinners (the high honor seat Gwen was to find out) often struggled to make conversation with him. Gwen said he ate silently and gave very short answers, arising abruptly to signal the end of the meal and to be escorted out, signaling the end of the event.

I walked in circles around my hotel bedroom. Wondering whether I should ask for time to consider my response, I churned over all the reasons I should accept or reject the offer. It was midnight when I called Gwen in Connecticut and once again shocked her with my decision.

Remembering the previous occasion when I'd had both good and bad news to impart, I said, "This time I have only good news for you. I have been offered the vice chancellorship. The job will be mine if I say 'yes' tomorrow morning. I need to sleep soundly tonight so that I can do well with the meetings with the staff and students. I have decided that I should accept the offer."

As usual, Gwen was cautiously enthusiastic and supported my decision. Thus began my second career. This time I was to be in a completely different role to my industrial self. I was to be the academic and administrative CEO in academia. My earlier four years as a transient academic served me well. I felt that I was reasonably knowledgeable about the academic world through my experience in establishing a department at CUHK, and that I had witnessed the fruits of the labor of everyone in the department, specifically the fifteen self-assured and knowledgeable graduates four years later. Of those first graduates, one woman succeeded in receiving her doctorate and joined the teaching staff of an American Midwestern university; another ended up pursuing a successful career in telecommunications. A majority of them made their mark on the community in one way or another.

Despite the confrontational manner of the staff representatives and the student union leaders during the meeting that I was introduced to them as the vice chancellor-designate, I did not believe that they were attacking me personally. They simply were trying to voice their unease about how the university council had excluded the staff and student representatives in the selection committee. I emphasized the need to have a strong research culture to raise academic standards. I remembered what the new Canadian VC had said about having a specific target as the way to add excellence to a university. I needed to deflect the line of questions calling into doubt my administrative ability to govern the university.

I said purposefully, "Each of you, the academic staff of this university, is proud of your own achievements in the field in which you specialize. I shall do my utmost to create the space for you to grow to greater eminence in your contributions to your field and in your care and nurturing of your students."

On reflection, I was putting myself forward as an idealist and could legitimately have been criticized as being too naïve. Unfortunately for me, this is my nature. My own defense of my naïveté is that I am simply too sincere. As events turned out over the following nine long years, I was never attracted to the intrigues of the university and its people. I wore my armor well. My strict discipline of adhering to honesty has always allowed me to sleep well and to speak freely.

I then flew back to Europe to attend the rest of my meetings in a rather unsettled state of mind, anxious to rejoin my family. To my horror, when I went back to my office at ATC in Shelton, I found the news of my appointment had preceded me. The overnight editions of the Hong Kong Chinese newspapers sent to California and elsewhere in the States had splashed the news all over the front pages. The news had reverberated in the Chinese community from coast to coast. I had not yet notified my boss of my intended resignation! I faced my boss in embarrassment but found him delighted for my opportunity: "Congratulations. We are very sorry to lose your services, but you have greener pastures ahead of you. ITT had other plans for you as it goes through big changes, and we shall miss you. Good luck to you!"

The interim year as vice chancellor-designate was an exciting time for Gwen and me. A new job, with a six-year tenure assured, meant that we could plan for the use of our time accordingly. Our children had both graduated from universities and were already productive persons in the labor force. After nearly thirty-five years of continuous service with ITT, inclusive of the four years of leave in Hong Kong, we felt a pang of sentimental pain at the severance. The company had been good to me. It was the top brass of ITT that had set the wheels in motion through their persistent lobbying of the various prominent prize award committees; the results of this support were the first of what was to become an avalanche of international and domestic awards, of prizes in recognition of my pioneering work in fiber optics.

One of the first major awards in America was from the Franklin Institute in Philadelphia. At the dinner to celebrate, I remember the jocular remark from Bill, one of the top ITT brass. He was then technical director of research for Europe: "This is just the first one, young man. You had better get used to all the hoopla. After it starts there will be many, many more!"

At about this time, the negotiations between ITT and Alcatel reached a mutual agreement. ITT was selling its entire telecommunications business to Alcatel, including me. In view of this we felt it timely to take an early retirement and to leave for new pastures. Alcatel's new management did offer me a post as the research laboratory director of the U.S. research arm. I had to disappoint the Alcatel management by declining that offer. After winding down the project that I was building, I left ITT with a pension and a clock to remember my time with ITT. It had been a unique experience. My duties at ITT/Alcatel were winding down, which allowed me time to supervise unofficially the affairs of the university.

My colleagues at ITT held a farewell dinner for me. I admit to being tearful. Gwen held her breath, fearful that I would break down completely. I had to keep swallowing hard.

While the outgoing vice chancellor continued to make the day-to-day decisions, he made a point of consulting with me when his decisions would open new directions in the progress of the university. I was most grateful for this. It was a period of adjustment and learning for me. Several of the senior staff from the university came to visit me from time to time to brief me more fully on future related matters. However, the chairman of the council did not want me to prematurely learn about one of the most knotty problems confronting the university. The chairman apparently had been advised that the government would like to line up the tertiary education into a uniform system after my taking office. The chairman hinted that I should think through how the matter could be handled. He advised me to be sensitive to the strong sentiments of the staff and students on this matter.

Between the date of my retirement from ITT/Alcatel and starting date at The Chinese University of Hong Kong, I had approximately three months of time to put my affairs in order. It so happened that an interesting offer came from Bellcore, a relatively new entity established after the Supreme Court in the U.S. ordered the splitting of AT&T into seven independent operating companies (known affectionately as baby Bells) to prevent AT&T from holding a monopoly over the industry. Bellcore was the joint research arm of the seven baby Bells. The director in charge of the entire research division approached me to see whether I would work as a consultant to him. He apparently wanted an outsider's opinion. I had a long association with Bell Laboratories and many of the staff in their research and development divisions, but as a competitor I'd never had the chance to really get to know the organization. I was delighted to take on this task. The research director offered me the position since he knew that I had left ITT and that I was to be the head of a university soon.

There was no conflict of interest that he needed to worry about. Every-one knew that the members of the technical staff from Bell Laboratories were all top-flight researchers, and that the management relied on their individual innovative and inventive talents to achieve results. Now as an insider, I could have a clear understanding of the research culture and methodology that made Bell Laboratories the most respected research organization in the United States.

As my task of critically evaluating the status of a large research divi-sion spanned over a three-month period, I had the opportunity to work with many of the researchers to get a sense of their aspirations and frus-trations, to witness their approaches to solving problems, and to analyze a range of management issues. This added another dimension to my ex-perience base and helped me in later years to become a technology con-sultant. My major conclusion was, in hindsight, an obvious one from a management point of view. When the research director assembled his team, he carefully handpicked the up-and-coming young talent and a number of experienced leaders; all of them were the best in their class. Later, when I presented my evaluation, I concluded that such a group of young lions would come up with great new projects with far-reaching prospects for growth. This was indeed the case. As a consequence, almost all of the innovative projects were maturing to the point that major ex-pansion needed to be implemented. Unfortunately, management could not get the increased funding necessary to sustain all of the great proj-ects; hence, many of the projects had to be curtailed. A drop in morale started in earnest, which was a direct consequence of having too many innovative and prolific performers, a problem of having too much of a good thing. I left Bellcore with an improved working relationship with the Research Division and many new friends. A few of them later joined CUHK as part of the staff in the Information Engineering Department established a few years later. These experienced staff members signifi-cantly improved the quality of the department's program and raised its status to world-class levels.

In 1985 I had spent a year working at Siemens in Stuttgart. My son, Simon, who was at his first job, stayed in our home in Trumbull and commuted from there to his nearby office. Amanda was working at Bell Labs in New Jersey and had her own place there. Returning in the winter of 1985, the first task had been to arrange surgery for Gwen. She had developed ovarian cysts. We spent that Christmas at home as a family, together after long distances apart. This was to become routine in the years ahead—long distances of separation.

As Gwen recovered from her surgery, she began to look into courses at the nearby Fairfield College for the spring term and Simon was being encouraged by his peers to get his Master's. He ended up enrolling in Brown University's Computer Engineering Department in 1986, and we will always be grateful to his colleagues, because this particular advice to continue his education could never have come from his parents.

With all of this happening, it was comfortable being back in our own home and returning to a normal, familiar routine.

The CUHK decision was nerve racking. I wondered if I had made the right move. Like all major decisions in life, one is beset by doubts afterwards. However, the follow-ups and continued contact made by the university via phone, fax, mail, and personal visitations by senior staff to keep me informed made me feel more at ease. I was seamlessly being drawn in over the months before I actually took office.

It was a bright autumn day when I landed at Kai Tak Airport as the official third vice chancellor of CUHK. Gwen and I had a comfortable overnight flight and were ready to meet the welcoming group consisting of pro-vice chancellors, registrar, university secretary, bursar, and university librarian. This was the type of formality still practiced in colonial Hong Kong. We alighted from the first-class section of the plane onto the tarmac and were greeted by a protocol person who guided us into a waiting car parked beside the plane. Royalty must have arrived in the same manner; all we lacked was a red carpet. We grinned at the bemused passengers who had to line up to wait to board the bus to the terminal. We were driven to the VIP lounge where the welcoming party had gathered. It was quite an occasion, especially as it was very early in the morning. We sat in a well-furnished room with sufficient sofas to hold all of us as tea was served, while our passport formalities were completed for us. Thus began my tenure as vice chancellor. The chauffer and car were waiting for us with our luggage already packed into the trunk when we emerged from the VIP lounge. The news media was also waiting. Cameras flashed as they snapped pictures of our arrival in Hong Kong. We had turned yet another new page in our life.

The vice chancellor's lodge was located on a hill opposite the main campus. It has the longest driveway in Hong Kong to a private house. At the top of the drive, the guard opened the big iron gates when he spotted the official car. Our major domo, the supervisory clerk who was assigned to us for household purchases and general management of the house, was there to greet us. We realized only later the inconvenience of this arrangement, but at that moment we simply took everything in stride and with a sense of bewilderment at being thrust into such a lifestyle.

The only house on the hillside, the vice chancellor's lodge, stood on a stretch of flat ground halfway up a well-wooded hill. From the roadside, the entrance of the driveway could easily be missed among the trees. A small, low, stone pillar with the words "Han Yuen" was the only sign of a dwelling. The house, based on a 1950s California design, was built in the 1960s. The downstairs was designed to accommodate sixteen or so persons for dinner with an adjoining spacious lounge for a multitude of functions. We often invited thirty to fifty staff and guests or an entire department of staff and spouses for a variety of functions. With the buffet-style snack food in the dining area and with the open doors leading onto the spacious garden, the guests could wander and enjoy a bird's-eye view of the entire campus. The center of attraction in the garden was the fishpond. Feeding time for the fish was a spectacular show. Red, gold, white, and speckled species of all sizes presented a feast for the eyes. The Japanese Consulate had given these beautiful species of carp to the university at the time of its inauguration.

Our arrival stirred a flurry of activities. The CUHK staff was everywhere helping us settle in. To their credit, they quickly left us to get ourselves organized. Some temporary staff remained behind to deal with cooking and other household chores. As we walked around the grounds, we immediately appreciated the tranquility of the place and the freshness of the air. We soon retreated to our bedroom to combat our jetlag. When I woke up after a couple of hours' sleep, I felt relaxed and at home.

If I had not been at least partially prepared for a whole year in advance to understand the university and its mission, and to know how to deal with its administrative processes, I would have been overwhelmed with the daily routines. It would have been too late to formulate what this university should aspire to and to devise a plan of action to rally the troops and achieve these targets.

My clarion call was to raise the quality of research to a standard comparable to those achieved in well-known universities around the world. I repeatedly reminded the academic staff that they were participating in such research before they returned to Hong Kong after finishing their higher degrees and after years of working at established universities. Surely we could show the world that we can be as productive and thorough in our academic pursuits. I promised to support their efforts with needed equipment and resources through raising more money from donors. I was convinced that good research led to good teaching. Good teachers could not remain excellent if they lagged behind in their pursuit of new knowledge. I would quote Confucius, "To revisit an old subject is a sure way of finding something new," to be a truism for advancing knowledge. I was convinced that having top-rate people is the only means to achieve excellence.

If there is a common streak of stubbornness in academic teachers, it is in their confidence in themselves. No academic can rise to the top without such a tendency. For it is that self-confidence, coupled with a clear sense of reasoning, which makes a convincing teacher. Such a person will accept his or her mistakes but never concede their beliefs until proven otherwise. I love to face such staff and relish their sense of knowledge. Even if they begin in an unreasonable manner, they will yield with reasoned persuasion.

I believe that the secret weapon of a vice chancellor is the ability to understand the aspirations of the academic staff and to be able to communicate with people effectively. Unfortunately, some scholars lose self-motivation and have no impulse to progress. They soon become dead wood. The worst ones are those who work hard at demoralizing their productive colleagues through promoting nondeserving staff. An excellent department can be mortally injured through voting a mediocre person to be department chair. Motivating the best performers is the only antidote.

Over my nine years serving as the vice chancellor of CUHK, I learned that I had two invaluable advantages. First, I was able to push excellence, because I had sufficient first-hand experience working with top-rate people. My own reputation also helped to some extent. Second, I was able to stay neutral, and could be brutally honest. These characteristics helped me to resolve important matters without antagonizing either the majority or the minority group.

At my retirement party, the farewell was a joyous occasion. We had overcome many obstacles in seeing the emergence of CUHK as a world-class institution. As the chairperson of the University Grants Committee (UGC) said, "CUHK was a university in Hong Kong but now CUHK is an internationally known university of repute in Hong Kong." The credit goes to the entire staff, especially the productive academic staff. Of course, I happened to be coming in at a stage of the development of CUHK that made my attempt at raising the quality of CUHK to an international level of excellence possible.

The Four- to Three-year Battle: A Battle Lost but Won
I was not well prepared to face the task of converting a proven four-year academic program. I stared disbelievingly at the request from the UGC that CUHK would receive the same funding that was accorded to the University of Hong Kong, where the academic program was based on the completion of a three-year program, with students needing A-level Examination qualifications from secondary schools.

This seemed absolutely ridiculous to me. Why should the UGC, at the request of the government of Hong Kong, simply make the funding methodology of the tertiary education sector uniform across all tertiary institutions? My initial reaction was logical, which was that it was simply a stupid bureaucratic decision that ignored and closed the door on a heterogeneous tertiary system of education, which should be the aspiration of any large metropolitan region. If the government wanted to expand the number of graduates from the tertiary sector, a three-year program would be less expensive.

Forcing CUHK to conform most likely originated from the secondary school sector. I found that one of the objections to the four-year system (this from a headmistress of one of the leading schools for girls) was that they wanted to retain their girls for one extra year in the secondary schools. They were looked upon as role models for the younger girls, and they would be helpful in playing a leading role in school functions and programs.

But as far as I was concerned, I wanted the universities to become more flexible, rather than to have all of them molded into the same form and function. My reaction to the UGC request was to find ways and means to rescue the university from this catastrophic decision. I wanted to preserve the structure that would allow universities and college systems to be flexible.

A simple solution was to improve the flexibility of the university as a primary means of allowing CUHK to move towards improved quality and excellence, at least in our traditionally strong areas and new initiatives. The solution was to move our programs in all areas on to a credit unit basis. The curriculum in a program leading to a degree must contain courses, each of which is worth 1–3 credit units, with a total of 120 credits necessary for graduation. I preferred this approach since it had been practiced in the U.S. successfully for a significant number of years at a wide variety of universities and other tertiary institutions. The flexibility of this system was high and could be implemented easily.

This move solved many urgent problems that were facing us, and provided the flexibility that we would need to achieve quality and excellence. Resource availability was critical, as CUHK had, since its inception, been funded at a lower level as compared with the University of Hong Kong. I could say to my colleagues, quite truthfully, that my recommendation to the UGC's proposal was for it to be accepted without reservation.

I told the staff and students of the university that we were not abandoning the four-year program. We simply said that a student entering this university could pursue their studies at any pace and graduate when the student had accumulated 120 credits. This could be done in three years on average, if at the time of entry they had successfully passed the A-level Examinations and languages requirements. Such entrants would be given 21 credit units in recognition of the A-level qualification. In one fell stroke, CUHK could proudly retain their historical tradition. At a later point, CUHK could move to accept students with community college qualifications, with transferable credits from other universities and from the future changes with little difficulties. In 2002, the government proposed to move to a six-year primary plus a six-year secondary school system. The universities would structure their credit unit courses accordingly and allow students to graduate, on average, in four years. In hindsight, the detour was unnecessary and wasted the energy of many people. I do hope that those responsible for the current reforms see the need for flexibility. The move to add privately funded schools to Hong Kong is the first in a series of moves that could allow Hong Kong to develop the necessary human resources for the coming Knowledge Era.

I was not satisfied with the way that CUHK had been forced to accept the proposed three-year system of university education. It appeared to me to be a backward move, especially when the entire university educational process was under considerable strain. It could not accommodate courses that should have greater flexibility, so as to tailor the courses for multidisciplinary subjects with flexible years for graduation. At CUHK, I had a tough time convincing all of my colleagues that we had not abandoned our four-year principle. In addition, I was facing structural and administrative changes that I envisaged to be necessary for the University to attain goals of quality and excellence.

The Tiananmen Ripple: Hong Kong Affairs Consultant

Nineteen eighty-nine was a turbulent year in China. A peaceful and orderly cry for an acceleration of the democratic process by the students escalated throughout the year. Whether the demonstrators really understood the true meaning of democracy was difficult to estimate, but the movement intensified, possibly through mass enthusiasm, or with ulterior political motives internally or externally exerted. It reached a crescendo when a large number of students and others from laborers to academics, took over Tiananmen Square in front of the Old Palace in Beijing. After vacillating for a few days, the government issued an order to clear the square, by force if the students refused to leave the area.

The CUHK student union had sent a contingent to lend their support to the movement a few days earlier. They were trapped when the tanks moved into the square. I was not in Hong Kong on that day, but I watched the reports on U.S. TV. Later reports indicated that a majority of the casualties were on adjacent roads leading to the square. Several days after the event, all the CUHK students returned safely to Hong Kong.

It was a harrowing time for the people of Hong Kong and, for that matter, for people all over the world. Hong Kong citizens turned out in droves. Over one million marched to a candlelight vigil at Victoria Park in Hong Kong in support of the students' expression of desire for a country without corruption, with respect for human rights, and against dictatorship. It sent a clear message that the people of China, including Hong Kong, wished to see the government of China do better.

There was a strong reaction from students in Hong Kong, and at CUHK in particular, to the Tiananmen Incident. The ripple effect throughout the CUHK students did not fade quickly. I was invited to be one of the "Hong Kong Advisors" to China, in response to the Incident. The Hong Kong advisors were to be a channel for opinions from people of different walks of life in Hong Kong.

I consulted with my senior colleagues and Gwen before accepting the invitation. Almost everyone, including the man who was destined to succeed me, suggested that I not take up the post. I told everyone that a channel of communication to the government of China was a conduit to express my views plainly and honestly. It felt unlikely that I would become, as students later expressed their concerns, a trumpeter of government propaganda.

The students demanded a public debate with me in an open forum. At the appointed time, I faced a large crowd of students, staff, and reporters.

I maintained my conviction that the open communication of views was essential for achieving truly democratic means to gain a common understanding, even if minority and majority extremes still held their same views. If we are responsible members of society, we surely should be able to see the many alternatives and then proceed towards the best solutions. The students took an aggressive stand throughout the debate. They apparently could not accept the fact that the Chinese leadership would listen to, or pay attention to, the opinions of the people.

Establishing the New Engineering Faculty

Reacting to the Tiananmen Incident, the government of Hong Kong took a major step in altering its policy of preparing for Hong Kong to return to Chinese sovereignty. The Governor of Hong Kong made a significant move designed to ensure the people of Hong Kong that the British government was eager to assist the people of Hong Kong, should China be foolish enough to renege on the Sino-British agreement of establishing a special administrative region of China, where the political and economic systems in Hong Kong would continue to remain as is for fifty years after the change of sovereignty in 1997; that Hong Kong would be governed under its own law, named the "Basic Law of Hong Kong," as previously agreed in 1984 between Britain and China. Britain would immediately grant qualified Hong Kong residents a special passport that would allow these residents British nationality and right of abode in the UK. The qualification is based on their permanent residency status, skills, and wealth requirements. This gave rise to a hot debate in the British Parliament over the responsibility of Britain for the people of Hong Kong who were colonial British subjects. However, opinions were divided in Britain whether the means testing meant that Britain was not living up to her responsibility for her colonial people. The Home Office was strongly against accepting everyone from Hong Kong, but agreed to issue a special B&O passport for Hong Kong residents to facilitate their travel abroad after 1997.

Simultaneously, the governor announced two major initiatives. One was to expand the tertiary education system by 50 percent within ten years, and the other was to build a new airport in Chap Lap Kok by reclaiming land adjacent to the tiny island of that name. This airport was estimated to cost around half a trillion Hong Kong dollars, roughly US$60 billion, and would serve as a hub for the region. In addition, two more container ports were approved for construction in addition to the seven container ports already operating. This would make Hong Kong the most complete and the largest transportation center in the entire area.

As far as I was concerned, the expansion of the tertiary sector came at just the right time. I was given more means to expand CUHK, which allowed me to compete on a more even basis with HKUST, the new technology institute. As part of this growth, CUHK proposed to start a new Faculty of Engineering, as well as to expand and to strengthen the existing departments. The Engineering Faculty was to focus on electronics engineering and computer science disciplines. The basic knowledge areas were electronics, computers, systems, and materials.

Hong Kong's stability was improving. The bidirectional investments between Hong Kong and China were active. Money from investors was increasing, and the GDP was steadily increasing. At the same time, a temporary downturn in the U.S. economy made our recruitment of first-rate staff possible and affordable. The Electronics Department and the Computer Science Department were placed within the Faculty of Engineering. The Information Engineering Department was the first new department to be established. We were lucky to have been able to persuade one of the best engineers from Columbia University to be the founding dean of the new faculty. The Information Engineering Department was unusually lucky in being able to attract people with both academic and industrial experience. They were mostly graduates from excellent universities and had worked for a significant number of years at renowned research and development establishments in the best-known communications and computer industries. I might have had some influence in attracting them to CUHK. Under the skillful hand of the dean, the Faculty of Engineering ranked well in comparison with equivalent university departments worldwide. This provided me, as the vice chancellor, with a comparative yardstick with which to encourage other departments to progress. Over the decade, all the departments improved significantly.

What conclusion can be drawn from the experience of starting a new faculty? The first and foremost requirement is the quality of the people. I do not believe that we have explored the synergy between people sufficiently. Through the examples set by great science- and technology-oriented research universities—such as Caltech, Berkeley, Stanford, and MIT—we showed that each professor with a group of post-doctoral researchers could achieve great things. It is timely to compare it with industrial-scale research where larger teams with several leading principal leaders covering a wider range of disciplines act as a guide. There is a role for the universities to fulfill, since many industrial concerns realize the effectiveness of research at the universities, if properly organized. Industries are already outsourcing such tasks to universities. The Area of Excellence programs recently introduced by UGC, if properly nurtured, could set new standards and rates of achievement in terms of knowledge and innovation. I believe that the second important requirement is to create the right environment within the universities, specifically aimed at performing contract research. This move would require the structure of the universities to be substantially modified so that the right talent can be grouped over several years to work on assigned projects and to utilize other research tools to advance basic research. The environment so described is alien to most universities and must be developed accordingly. Before we can run we should walk in that direction by first encouraging multidisciplined projects to take a lead position within a university or a group of universities.

The structure of the university supports the efforts of the academic and administrative staff so that the task of the university can be accomplished efficiently and so that each person can derive different but complete satisfaction. This is, of course, the ideal. It is the vice chancellor's responsibility to deal with support efforts. The VC cannot hold anyone's hand and still perform his or her numerous tasks. The vice chancellor must maintain external relationships with the community, the government, academic contacts throughout the world, and individual staff throughout the university. The constant administrative issues and academic directions require a great deal of time. Communication with staff and students takes yet another considerable percentage, and on top of that are the new initiatives.

I faced a number of structural reforms that required addressing, including leave entitlement for academic and administrative staff, automatic salary increments related to the superannuation scheme, criteria for the dismissal of nonperforming staff, intellectual property rights and outside practice, and budget allocation.

When the university started in the 1960s, the need for upgrading academic staff's knowledge was very important. In response to this situation the university council cast in concrete that all academic and senior administrative staff were entitled to two months' leave every year and could aggregate the leaves such that in the sixth year the person could take one year's leave. The purpose behind this was to refresh and expand knowledge. The paucity of research resources in Hong Kong was the primary reason for staff being given leave to keep abreast of the person's field of expertise. For some unknown reason, the administrative personnel had the same privilege.

By 1987 I realized that hardly anyone ever took this one-year leave, and that most of the staff were instead holding on to their entitlement. In this situation, the staff could take the entitlement at retirement and then collect an extra year of pay. Since inflation was continuous, the person's entitlement easily could double every seven years, or even more quickly. One year the salary increase was 19 percent. There were several major problems with this scheme, with the primary one being that more than 50 percent of the staff was holding on to their leave.

The first problem was that if all the staff entitled to leave decided to take it in the same year, the university would have had a serious cash flow problem, as well as a severe staff shortage. The council rules were such that I found only one way to deal with this situation. I submitted a change proposal for the council to approve that all new staff must accept a revised leave arrangement. The maximum accumulation of leave was reduced to six months, which was to be taken after giving a nonchangeable start date. The existing staff could opt to change to the new scheme.

The academic and administrative staff were given equivalent grades pegged to the master scale of the civil service pay scale. The civil service pay scale was established in the colonial days when senior ranking staff were invariably sent over from the UK. Their pay included a so-called hardship compensation, and generous home leaves. When the university changed to an equivalent grade system, there was a clause attached stating that the perks must not be better than that of the equivalent grade of a civil servant. By the 1970s Hong Kong had set up an anticorruption bureau—the Independent Commission Against Corruption. This bureau has the power to deal with anyone found to be "corrupt," even if the person is the governor. For this bureau to be effective, it introduced a new pay scale for all civil servants such that they were paid at a level equivalent to or slightly higher than the private sector. This worked well. However, by the 1990s the salaries were rising very fast due to the pegging of the Hong Kong dollar to the U.S. dollar at a fixed rate of 7.8 to 1, with very small fluctuations permitted. Inflation in the United States accelerated the upward spiral of salaries in Hong Kong since Hong Kong civil service salaries were paid with an annual adjustment that included a cost of living increase based on inflation, as well as an incremental increase based on the automatic increase of salary to reflect an additional year of experience; except in cases where someone in the civil service had reached the top of the pay scale.

The impact was felt acutely throughout the universities because most of them operated a superannuation scheme that favored the employee when inflation occurs. After many years of inflation in Hong Kong, its GDP climbed to the territories of the wealthiest nations. Hong Kong at one time was ranked near the middle of the top ten countries. CUHK's superannuation reserve faced a fiscal deficit on an actuary computation basis. With the help of the council chairman and one council member, we devised a package so that the superannuation fund was limited and would remain solvent, irrespective of pressure from inflation or deflation.

When a staff member was found to be nonperforming, the council ordinance has only one condition under which to dismiss the staff, namely insanity or gross negligence of duty. At CUHK there was an additional condition for dismissal. This applied to any staff over 55 years of age. During my nine years of service, I got as far as identifying only one possible candidate to dismiss, but I was not in a position to pursue the case during my final year of service. In the real situation, staff can delay dismissal by suing the university for discriminatory treatment or false accusation. The case could drag on for years. The only truly effective way is not to renew the appointment at the age of 60. I was forced to accept that a small amount of dead wood could never be eliminated. The only remaining option was to appeal to the ego and encourage the person to do something constructive or do nothing at all and quietly disappear from the scene.

In science and engineering and other applied areas, incentives for staff to do outside practice or to claim intellectual rights could readily be found. In arts, languages, and literature, similar incentives are few and far between. My approach to the issues of outside practice and intellectual rights was shaped by my experience when I was the chairman of the Electronics Department.

I said to my colleagues in the department at the time: "I find that the work I have to do in my role can be divided into four parts. They are: administration, teaching, advising students, and research. Everyone must assign a percentage of their working hours to each of these four tasks. I also realize that each of these tasks can easily absorb all of my available working hours. Each of us may assign different percentages of time to each of the tasks according to one's own preference, but there might be some give and take necessary to ensure that all of the work is accomplished with the available staff."

Whether we should allow outside practice or work towards the ownership of IPR (Intellectual Property Rights) are debatable points. Such activities could help a person's own research or acquire consultancy skills, and both activities could allow a person to earn extra income. Should the university allow staff to engage in such extra work for financial gain?

If we apply the same rule to literary subjects, such opportunities appear in a different way. An artist or a writer could sell their class work by publishing their paintings or essays for profit. This is the dual use of intellectual output at no one's expense.

My belief is a pragmatic one. I prefer to rely on the sense of responsibility of the staff concerned. A responsible and motivated staff member will pursue excellence in one or more of the roles of an academic. Whether a person should be allowed to use 20 percent of the available time to pursue outside practice is not the correct question. It is better to use the American method of paying staff nine months' salary and then allowing staff to seek additional income over the remaining three months.

It is not wise for universities to attempt to acquire IPR as a business or to set up special outside practice opportunities within the university unless additional staff is recruited. Universities should not mix their mission with profit-making business motives. Town and gown can and should work together, but each side must maintain a clear view of their primary mission.

Nine years at the helm of a university was the best learning experience I had throughout my entire working career. I came across people from all walks of life and across the entire age spectrum. I loved the breadth of knowledge and the vastly different ways that people expressed their views. I could not have been exposed to this range of influences anywhere in the world, except at similar universities with a full range of disciplines. I am extremely thankful for my time as vice chancellor, which improved my ability to integrate and better appreciate the connectivity of thinking. My perceptions are not always necessarily correct, but I have become convinced that it is generally best to think simply. Complexity solves nothing.

A RANGE OF PROFESSOR KAO'S AWARDS

All My Prizes

One of my early prizes for work in the field of optics came when I was living in Roanoke. The local newspaper ran the story, which is how my next-door neighbor learned about the award: "I saw in the newspaper you're getting a Steuben bowl! Wow, I can't wait to see it. It'll be beautiful."

The Steuben factory is famous for its cut crystal glass—they produce large plates and dishes, bowls, and ornamental items of splendid designs, some with etched artistic details on them. They are usually very costly items. I was not aware of this at the time, so I was a little bewildered by my neighbor's excitement. It happened that the date of the Morey Award ceremony in 1976 clashed with an important overseas trip for me that could not be rescheduled. An emissary from work attended the ceremony on my behalf and I didn't pick up the bowl from him for some time.

My neighbor kept badgering me about it. Eventually the bowl arrived. It was a very simple plain round bowl, about five inches tall and eight inches in diameter—thick and heavy, especially at the base. My name and the date with the donor award were engraved on the side—quite weighty but not a particularly imposing object.

When my neighbor saw it, his jaw dropped in disappointment. He had imagined something much more elaborate.

The first prize that came with a monetary reward was the Rank Prize in 1978 from the UK. This prize was shared with my ex-colleague, George, who had been a co-author of the 1966 paper. It was a lot of money to me then. I didn't know whether to spend it or save it, so I did both.

After a few years in Roanoke, the family was feeling more settled and getting used to life in America. We were finding the vast countryside excellent for hiking; exploring the National Parks on the Appalachian Trail became a favorite pastime.

Smith Mountain Lake is the local getaway for weekends in that part of Virginia. The shoreline around it is about two hundred miles long and consists of mainly virgin woodlands with many intricate coves and hidden inlets. It was a developing recreational area, though already some magnificent holiday homes with boat docks had been built along the shore. Speedboats careened on the waters pulling water skiers.

Eventually I succumbed to the temptation and bought a piece of land, six heavily wooded acres on a lake inlet at the mouth of a tiny creek. Often only a trickle of water ran down from some hidden spring lost to sight in the scrub and trees. The main lake could be seen in the near distance. When the water was high it was easy to access the lake, though sometimes, at low water we faced a mud-bed. The undeveloped land was inexpensive, especially as it had no direct frontage to the lake.

On weekends we went there to putter around and to commune with nature. I learned how to cut down a tree with my chain saw. As the shore was so wooded, I needed to clear it to have some open land to use for camping or for cookouts. In the beginning I was not a very expert woodsman. Having tackled first the smaller trees so as to feel less nervous handling the noisy chainsaw, I quickly became more ambitious.

By the end of my first year as a landowner, I had built a wooden platform on which to camp and leave belongings; also a usable well-built floating boat dock from which we could sit, fish, or swim. The boat dock was anchored by ropes to two nearby trees. One night of terrific storms ripped it all loose. The whole contraption sailed off and disappeared. Before that we had bought a small second-hand boat, had learned to water ski, and unsuccessfully tried to fish out in the larger lake. We were able to go boating and to remember the landmarks around the lake. It was easy to get lost in the small coves as they all looked the same and our friends had their holiday homes tucked away out of sight in those inlets. During that same storm, the boat sank. Fortunately it was tied with a more secure knot to a sturdy tree. Several strong friends came to my assistance so we could raise the boat. It survived the drowning, but a lot of elbow grease was needed to get it into clean and dry condition.

It had always been a dream of ours that one day we would have a small holiday home built on this land. Now I had the money to make this dream true.

Gwen had been looking at log cabin packages and they looked easy to assemble, like fitting a jigsaw puzzle together. That was what I ended up spending the prize money on.

First we had to get a road put in, choose where we would site the cabin, and have a foundation dug and a cement slab set in. A septic tank for sewage had to be laid and, of most importance, we had to find a clean water supply.

The well digger actually did a divining search with a willow tree fork. Holding the branch in front of him, he walked through the brush and trees to a spot on higher ground and declared that the water could be found at this spot. His willow branch was quivering. Cynically, I wondered if he had decided in advance that it would be easier to drill nearer the road for his machinery to get in. Water was found at a depth of 250 feet, which made it a costly well. Gwen did most of the organizing with the contractors and as part of this, was the first person in the family to get a speeding ticket.

The distance from our home to the lake was about 35 miles; the first part of the route was on highways with little traffic before it became small country roads. Early one morning Gwen was rushing to meet workmen there. The roads were empty and she stepped on the gas. She heard the wail of the police car. "Ma'am, you were doing over 75! Didn't you see me waiting at the gas station back there?"

Speeding vehicles were supposed to notice him and slow down voluntarily so as to avoid getting a ticket. Gwen, normally a stickler for rules and regulations, was too absorbed in thinking about time and being late. She gazed at the policeman blankly. He thought she did not understand and so she kept pretending not to understand. Slowly he enunciated the penalties.

"Well ma'am, if I booked you for over 75 mph, you would get two points on your driving record. Do you understand what that means?

Well, never mind if you don't! There are lots of rules here in this country. As this is your first violation, I'll be kind. I'll report it as only over 65 mph. You will get a summons and a fine to pay. Don't do it again. Have a good day."

On paper, a log house is easy to assemble, each log numbered and already notched to fit in with the next. In reality, my son and I could barely lift one log, let alone raise it into position. It was clear we had to hire a labor force. So Mike, a wiry outdoorsman who was an itinerant worker for the log cabin supplier, came to supervise the job together with one of his friends. It was a slow project, raising the logs to build the walls. It took more than one summer.

We celebrated the topping of the last piece of the roof and then I was on my own again to finish the interior. I hired a plumber to put in the water pipes, my daughter and I laid the electrical wiring, and Gwen and I sawed out the stair supports and runners to go up to the loft space. My son, not as enthused about the whole project, was a reluctant helper. Together we framed the house and did the drywall. Gwen passed by an old Victorian house in the downtown that was being renovated and noticed all the beautiful old oak flooring being thrown out. They were more than happy to give it away to her.

First, we nailed the oak planks in the hard way, hammering the nails at an angle; until we learned there was a gadget we could rent that held the nail ready at the required angle. All I had to do was raise a heavy mallet and just hit it once and the nail was driven in tight. The speed of laying the floor more than tripled. Gwen, ever the bargain hunter, found a lot of junk ceramic mosaic tile discarded at the back of a tiling shop and picked that up. We used these for the tiny bathroom floor and for the kitchen countertops. Our own unique and creative patterns resulted. As the winters were quite cold, we put in a bright orange stove with a long high chimney going up through the roof of the same color. The holiday home was slowly coming together and our comfort levels staying over the weekends were much improved.

There were still lots of finishing touches to be made but I got busier at the plant and I was away on more business trips, which meant that our weekends of relaxation became more rare. There also were other activities constantly happening—shopping at the mall, school functions, and social engagements. It was nice to sleep late on Sundays and an afternoon at the lake left us with insufficient time to accomplish much. We left before we could fully enjoy our retreat in the woods.

I kept the vacation home for a number of years after I left Roanoke, renting it out to different tenants. Eventually I realized there was no going back and it was sold. I wonder if it is still standing after a quarter of a century? I cannot remember its exact location anymore, tucked away in the backwoods of Virginia. I have many happy and wistful memories of my times at the lake. The knowledge gained with the home construction was invaluable and the family commitment to the project drew us all much closer together.

The small balance of money left over from the Rank Award went towards the children's college fund. Eventually these loans were to be repaid by each of them in the years following graduation. This was my contribution to teaching them fiscal responsibility; in this way they would better appreciate the actual cost of the education.

More awards in one form or another, a plaque here, a framed certificate and medal there, came my way. A very formal occasion occurred in 1977 as part of my receiving the Stuart Ballantine Medal from the Franklin Institute in Philadelphia. The award was jointly given to three of us—including my colleagues Dr. Maurer of Corning and Dr. McChesney at Bell Labs.

The awards ceremony was held at the Institute in front of a large and distinguished audience. My family was invited for the weekend events and we were put up in one of the top hotels. My evening suit, from my Hong Kong years, was brought out from mothballs to air and one was rented for Simon. The women of the family bought long dresses. It was a new and nervous experience for us all, from the grand hotel setting to the social formalities, and from the laudatory speeches to the formal grand dinner in the Great Hall of the Franklin Museum. I was told: "You had better get used to all this! This is just the beginning of all the hoopla. You will get many more prizes in the years to come. One leads to another!"

The headquarters staff had been pushing my name forward as their nominee unbeknownst to me. They were delighted with my results and I appreciated their years of support of my work. Bill's forecast was right. Without their initial push, the recognition might never have occurred and I am grateful to them all.

The Ericsson Prize in 1979 from Sweden was announced in the form of a telegram, followed by a formal letter. Those were the days when urgent news—good or bad—came in the form of a yellow envelope delivery from Western Union or the post office. My father sent one from England when my mother died in 1976.

Not knowing its contents, it was with shaking hands that I opened the envelope. My initial reaction was relief that it was not bad news. It took a while for the good news to sink in. The prize was a US$250,000, and it was to be shared with my good friend Dr. Maurer of Corning Glass Company.

My mothballed black tie and suit were not needed. It was to be a black-tailed suit, white-tie affair, and I was to be fitted out with it on arrival in their country. We were invited to a week-long celebration of the award in Stockholm in November of that year. The children, then at their universities, chose not to join us. We were flown first class by SAS, the Swedish airlines, and greeted by name as we boarded. The airhostesses beamed at us and congratulated me. I was a celebrity!

The hotel was a far cry from the days when I first visited New York on a business trip and stayed at the modest YMCA. We could see the royal palace across the water from our window. The luxury of crisp clean bed sheets and the beautifully laid-out breakfast buffets were an indulgence that we felt we could easily become used to. Unused to these facilities and frugal by nature, we were cautious with expenses, even though we did not need to pay for them ourselves. Gwen had strong convictions on this matter—if we would not frivolously spend our own money, then we should not do so with other people's money.

When I handed over the hotel bill for Ericsson to settle, the man looked at it with surprise: "That is a very sensible amount—very unlike the winner who ran up astronomical bills with expensive trans-Atlantic phone calls to his family every day! It nearly broke our budget."

The award ceremonies were modeled after those held for the Nobel. The same historic hall was used and King Carl XVI Gustaf of Sweden presided over it all. The Queen did not attend, as she was expecting her first child. A violin quintet played in the background as the distinguished guests from around the country were seated. The King, entering last, was seated in a chair at the front of the audience. He entered the hall with much fanfare. He was a young man about my height. I was glad he didn't tower over me like the majority of my American colleagues. It made me feel less nervous.

There were speeches and citations that gave a background of our respective work. Then it was our turn to give a reply. Gwen told me afterwards that my fists were clenched tightly as I stood in front of everyone, displaying my stress.

A full program of events had been organized for the week. There were many dinners besides the one on the evening of the awards ceremony. That particular event had been held in a large medieval hall with high arching roofs. Everyone was dressed very formally. There was another, smaller dinner that the King also attended. At the pre-dinner cocktail party, we found all the guests standing at one end of a long room and the King by himself at the other. Barbara, the wife of my fellow awardee, thought he looked very lonely and unhappy. "Come on," she said, dragging Gwen with her, "Let's go and talk to him. Everyone seems to be ignoring the poor man!"

They spoke with him in an informal American way, enquiring about his wife, and generally making small talk. Later that evening we were drilled in Swedish protocol. No one dares approach the King unless he approaches him or her first. He is a royal personage! Well, at least that explained why everyone was keeping their distance from him. Another rule is that when the King stands to leave after dinner, everyone must leave immediately, following him out. Gwen said she had a spoon of pudding halfway to her mouth when everyone stood up suddenly and started filing out.

During my conversation with the King he became quite animated, telling me about his yacht and asking if fiber optics could be used to improve the gyroscope on board his boat.

In 1988, almost a decade later, having been inducted into the Swedish Academy of Engineers (IVA), we attended the dinner in Stockholm for newly elected fellows, at which both the King and Queen were present. We stood in the receiving line and were introduced to them both. Now less gauche and more educated in the pomp and ceremony of public life, I bowed politely. "I have seen you before! We met at another event. Your face is familiar," the King said, smiling at me.

We talked of the past occasion and I was able to add that I was delighted to have the honor of meeting the Queen this time. King Carl seemed more relaxed now, too. Becoming a family man had made a difference. We had both mellowed with more experience under our belts.

One of the leading old families of Stockholm was the Wallenbergs. They hosted another of the dinners for the Ericsson Prize winners in one of their magnificent homes. Our Ericsson hosts arranged a day out in Drottningen, where the summer home of the royal family is located. Although it was a sunny and crisp autumn day, the frosty nip in the air signaled the coming winter. The gardens of the estate around the castle were large and lightly wooded. We were told that the Swedish princess liked to go riding through these grounds. A small theatre was built in the castle a century or so ago by one of the earlier royal residents. It had been restored to its former glory with painted panels on the walls. Concerts and other performances were being held there again in the summer for the local community to enjoy. It was a pity, we were told, that we could not go farther north to the Arctic Circle. The northern winter weather in November sadly ruled out that opportunity. Regardless, we had a wonderful week with royal treatment all around.

There were two medals given to me for the Ericsson Prize; one was a solid gold medal that I promptly stored in our bank box and the other was a copy made of a lesser metal. This copy was for showing off, so I took it with me in its red velvet box when I went off for my year in Germany in 1985. One day when I opened the box to show my medal to a friend, it was missing—gone. I can only assume that the cleaning lady who came in once a week to clean the rented house took it. She was a little deranged and we had had to fire her some weeks earlier. It was too late to confront her. Somewhere in a drawer in Germany, the worthless copy lies forgotten. Perhaps some day it will be found and I hope the finder will return it to me.

While I was in Germany, notification came to me of another award in a much more dramatic fashion. The method of communication was by long distance phone call from New York. There is a six-hour time difference between there and Stuttgart. The Marconi Prize selection committee was anxious to convey the news of their choice to the winner as soon as the meeting was over. They diligently located and called me. In a sleep-dazed state of mind, I picked up the instrument and grunted into the mouthpiece: "Uh—err. Yeh. Oh! Um. Thank you very much."

My eyes closed again as I went to sleep. Gwen was curious who had phoned in the middle of the night, and she kept nudging me awake. "I think she said the Marconi Prize!"

I turned over on my side and was out for the count again. Gwen, prodding me for more details, continued to disturb my sleep.

When I met the committee in New York, I was deeply apologetic about my graceless reception of the great news. Mrs. Braga, the daughter of Marconi, and I had a good laugh about it.

Mrs. Braga is the daughter from the first marriage of Marconi. The Italian family resulting from his second marriage is more prominent and Mrs. Braga was determined that her side not be pushed into obscurity. The fund she ran and managed was her means of maintaining this visiblity.

With her transcontinental connections in the U.S. and Italy, the award ceremonies took place in these two countries on a rotating basis. I was the eleventh winner in 1985, with the prize being awarded biannually. The 1985 award then was presented the following year in New York in the United Nations building, and the presenter was the fifth Secretary General of the UN, Mr. Javier Perez de Cuellar. In accepting the monetary prize, I also had the task of reporting on how I used the money at the following awards ceremony.

I had been giving a graduate course on fiber optics at Yale University in New Haven in the fall of 1984, and had repeated and refined the lectures as I delivered them again in 1986 and 1987. My lecture notes were growing to become an immense pile of papers. If I could only focus the piles of paper, it might be turned into a book. There were still relatively few resources for students on this emerging subject and I had already been approached to write a book by an institution in the UK. This was the project that I decided to focus on for the Marconi commission. In 1988, Peter Peregrinus Ltd. in the UK published my book *Optical Fibre*, under the auspices of the Institution of Electrical Engineers, IEE, as a textbook for students. It includes a wonderful preface written by Gwen, who spent many long hours editing my English.

The twelfth fellow was awarded his Marconi commission in Bologna, Italy in 1988, and it was there that I gave my report on the two years since receiving my award. Mrs. Braga made full use of her Italian connections and we had a wonderful time.

By that time I had already moved to Hong Kong to become the vice chancellor of CUHK. Simon had moved out and was at Brown University studying for his Master's. Amanda, who had joined AT&T after graduation from Duke University, had been sent by them to Berkeley to work on her Master's. The family was scattered around the world and the idea of a home to hold us all was a wistful dream unlikely ever to become a reality.

"The parents are not supposed to leave home, we are supposed to be the ones that do that!" had been my daughter's unhappy comment.

After leaving the house empty over a cold snowy winter we sold it in 1988. A small apartment was bought in New York to replace it.

The sale was made about the time of the Marconi date in Italy, so we worked our way to Rome via the U.S. and collected extra suitcases of belongings to take back to Hong Kong. The rest of the contents of the house were sold to friends, discarded, or shipped out to furnish the New York apartment.

We arrived in Rome by train with more luggage than we could handle ourselves. It took awhile to find a porter to get it all to the front of the station. The porter loaded the bags into the trunk of the first taxi and we gave him a tip in liras. Not understanding the value, we thought the equivalent amount in U.S. currency of a couple of dollars was sufficient. He did not agree. He spat on the ground and in voluble, incomprehensible Italian—probably cursing us at great length—reached into the trunk and threw our luggage onto the ground.

Another taxi driver, who had witnessed it all, ran over and drove us to our hotel. He would not accept a tip. I was very grateful to him for saving us. But what an introduction to the great city of Roma!

Mrs. Braga had tried to arrange an audience with the Pope; unfortunately the timing was wrong as another eminent statesman was in town visiting and he could not fit us in. Instead we were allowed a visit behind the scenes of the Vatican. It was a unique occasion. The Sistine Chapel was undergoing renovation and not open to the public. We went underground to the see the newly exposed ancient remains of Rome that had been discovered during recent renovation work. Our guide related to us how the holy relics of St. Peter were rediscovered during the excavations. They are now housed in a special cache below the ground level of the church and can be viewed from the top of the circular steps leading downwards to it. It was the duty of every newly appointed bishop to go down and pray in front of the relics and only on such occasions were the doors opened for this homage. As we toured further we marveled at the beauty of the newly restored paintings on the ceilings and walls of the other halls.

From Rome, our party was flown to Bologna in the private Lear jet of the President of Italy. Mrs. Braga had influential friends indeed. I had only flown once before in a private plane. That was when ITT flew their top management, accompanied by their wives, to an offsite meeting retreat on the exotic island of Lanzarote. That retreat, together with earlier ones held at the Greenbrier in West Virginia, had groomed me in the finer etiquette of formal dining and dignified public social life of the West; experiences that were to be vital in my subsequent exposure to royalty and leaders of world public affairs. In Bologna, its mayor welcomed us to his ancient city. He hosted a dinner to which all the local leading citizens were guests. One of the oldest universities in the world is in Bologna, its campus crammed with medieval marvels.

I have forgotten now who officiated at the Marconi prize presentation. All I can remember is that it was an Italian scientist. Italian is such a musical language that without any understanding at all, the many speeches soon lulled me into a drowsy state. The ceremonies were held in a building that appeared to have once been a church. It was on the other side of the market square from my hotel. Walking to it enabled me to pass through the older quarters of the city, down side streets that were now primarily retail spaces. I always find shopping areas of great interest and Gwen is always game for an adventure. I aimed in one direction, turning in and out of one narrow lane to another, stopping to look into the food shops, noting the prices and comparing them with those at home. The cheese shops were absolutely wonderful; more kinds of cheese than we had ever seen in our lives hung from hooks, and were piled on the shelves and stacked on the floors. The various dried hams smelled delicious.

We were also taken to visit the house where Marconi worked, which was a short bus ride away. It is now a museum in which his records and research materials are housed. In the well-kept grounds is a memorial to Marconi, as well as a larger-than-life bronze statue of him. It was from this house that Marconi made his first successful transmission over radio waves, talking to his assistant standing on a distant hill.

The first time we walked to the building on the other side of the market square was for the rehearsal of the awards ceremony. The second visit was for the actual event. The seats in the hall were now filled with local dignitaries and distinguished guests. One of the past awardees present was Dr. Inose from Japan. He came with his wife, and she and Gwen instantly struck up a friendship.

I gave my lecture and spoke about the history of how the book came to be published and gave a brief layman's version of how optics worked and the theory behind it. As it was given in English, and the audience's level of comprehension could not have been very high, I am afraid that it was not of great interest to most of the people present. Afterwards, the photographers had a field day. First the Marconi Fellows on their own, then with Mrs. Braga, then with her staff, and finally with the mayor. The combination of poses was seemingly endless. I had left Gwen talking with

Mrs. Inose. After some time I went to look for her and could not find her anywhere in the hall. Apparently she had tried to find me when she had gotten bored with all of the sitting around and waiting. She could not find me in any of the groups. Most of the local guests had left anyway, so she did, too. I was relieved to find her resting in the room when I returned. She had decided to wander the maze of side streets and eventually found her way back to the hotel.

Mrs. Braga had worked hard to raise funds and to obtain support from both Italian and American sources since the Marconi Fellowship had been established in the 1960s. Now that I was a Fellow, I was soon pulled in to serve on the selection committee. It was in this way that I got to know Mrs. Braga better. She was determined and extremely persuasive in her approach. It was difficult to refuse her requests. She oversaw the awards events with a firm and polished hand. She literally waved her hand and orders would be taken care of.

Gwen and I went to lunch with her at her home in northern New Jersey. Having lived there for decades, she had made the place into a very comfortable home. As it was an older property it retained its larger garden when other properties of the same vintage had been redeveloped into several homes on smaller plots. Over lunch she intimated that her half sister would gladly wrest control of the fund if given a chance. There was no way this would be allowed, so the half sister was never included in any of the events.

With the loss of Mrs. Braga at the helm (she died in the late 1990s), there was no one to take over. The Marconi Fellows have continued to assist in administering the fund, but her presence is sorely missed.

Japanese politeness is awesome in its intricacies, and I remembered how nerve-wracking it was to remember to make all of the correct bows. Dr. Inose was a kind guide. I remembered him from when we had first met in Japan in 1987, a year before. He had served on the selection board of the C&C Prize the year I was chosen as its winner. We were delighted to see each other again, and I remained in awe of his contributions to science.

As part of this award we had been invited to a week in Tokyo where we stayed in the grand Osaka Hotel. Wined and dined at expensive restaurants, and meeting with leading persons in the arena of politics and industry in the Japanese capital was quite an experience for an ordinary person like me. Afterwards we met the Inoses's on several other occasions. We exchanged greetings at Christmas over the subsequent decades. His annual card always had a nice photo of them both taken on annual vacations at many different and interesting places around the world. The year that the card failed to arrive in 2000 I had the distinct feeling that he had passed away. This sad news was confirmed later when Gwen wrote to them and Mrs. Inose replied with the news of his death.

But I am jumping too far ahead as I reminisce. Before this, I had the honor of being the recipient of the 1996 Japan Prize. This was a major award with serious prize money. The hope was that it would become, over time, equal in stature to the Nobel Prize. There is no category for engineering as a subject. The Nobel Prizes are given for discoveries and breakthroughs in the pure sciences. The Ericsson Prize hoped to catch the public's eye, too, but in all of the years since 1979, I do not remember ever seeing any news coverage about this award. I do hope the Japan Prize will become more internationally known, and not for personal reasons, but because engineering advances have made a great many differences to society and it is a profession that deserves to be more publicly recognized. Somehow a prize by any other name does not resonate with the people of the world in quite the same way as that of the Nobel Prize.

By the time I received the Japan award, I was about to retire from my post at the university. I was already set financially for the years ahead. I was no longer strapped for cash to pay for my children's education. They were looking after themselves and their own fortunes. So what should I do with the huge monetary award? I was the sole recipient. Gwen was distressed by the percentage that the U.S. government would demand in the form of tax. As American citizens, we were taxed on our worldwide income, regardless of its source, even if the income was for scientific awards paid out from foreign sources. It seemed unfair—but that is the tax code.

"I have no interest in a fur coat, elaborate jewels, or an expensive car and yacht, or indeed anything that flashy!" said Gwen. "Let's give it to a charity so that Uncle Sam won't pocket 40 percent!"

Gwen consulted a tax lawyer. It would have to be a charity registered in the United States. The lawyer suggested that instead of the lengthy time it would take to set up a charity that we could run ourselves, we should make use of my connections in the university community.

It sounded like a good idea to me. This is why all of the prize money went to Yale University. Although I am not an alumni of Yale, my connections with that institution run deep; from my teaching role there to my welcoming Yale-in-China staff to the Hong Kong campus during my tenure as vice chancellor. The objective of the fund was to aid students in their studies of East/West technology transfer. I hoped it would build bridges to deepen the understanding of Asian cultures in the West. I served on the selection committee of the fund to vet the applications from Yale students. After five years, the Charles Kao Fund was able to send about ten Yale students per year to the Far East during the summer months. Most of the students went during their senior year or in their early years of pursuing a higher degree. They were encouraged to add onsite experience as a way of furthering their individual projects. After a slow start, the fund and program are now quite well established.

One of the medals I did not manage to collect personally was the World Federation of Engineering Organizations Gold Medal. It was to be awarded at a ceremony in Bucharest. Gwen and I were already on the plane when the center of a typhoon hit Kai Tak Airport. The gale-force winds delayed our departure. We were worried enough at the time to de-plane, and it would be several years before the medal reached my hands at another engineering conference.

Another medal was awarded by the Royal Academy of Engineering in the UK. The ceremony took place in the Chinese Room in Buckingham Palace. Prince Phillip, who was the Patron of the Royal Academy, did the honor of presenting the medal to me. His Royal Highness asked me a few questions about technology and briefly explained to Gwen and myself and to the President of the Royal Academy of Engineering the origin of the Chinese Room. Prince Phillip was not as tall as I had expected.

Prince Charles, whom I met later in Hong Kong, was also not as tall as I had imagined. The press photographers must have worked hard to portray the Royals from the best possible angle.

The president of the National Academy of Engineering in the U.S., whom I knew and had entertained to dinner in Hong Kong a number of times, stopped by my office one day. I thought he was just visiting the area again. "Hello, how nice to hear from you. When did you fly in and how long will you be staying?"

I was surprised to learn that he was calling from America. He said he had some good news for me. I remembered then that my good friend Dr. Wing had been telling me over the past year or so that he was putting my name forward for the Charles Stark Draper Prize of the NAE. I was very pleased that finally I was being recognized in the U.S. for my own work. I was equally as pleased that the NAE was opting to give this honor to me, Bob Maurer, and John McChesney, as the co-recipients of the 1999 prize.

The awards ceremony was held in Washington D.C. in one of the State Department buildings. The scientific advisor to the President represented President Clinton in reading the presidential message of congratulations to the three recipients. There was a pre-announcement ceremony earlier to announce the award to the press and then a post-award ceremony in which each of the recipients gave a speech about his specific contributions. The latter event was held at Draper's Laboratory. Dr. Draper was a pioneer of navigational instrumentation that made the deployment of satellites possible.

Apart from the various medals and prizes, the honors I received included honorary degrees and professorships. The honorary professorships carry few defined duties, and are usually fulfilled by the person giving a lecture at the university. An adjunct professor serves as a supervisor to higher-degree students or teaching a class for a term. At the Yale University graduation of 1997, I was one of the honorary graduates, along with other recipients that included Federal Reserve Bank president Alan Greenspan and actress Julie Andrews. I happened to sit next to Greenspan and was able to chat with him as we were waiting for the ceremony to begin. Miss

Andrews was much taller than I'd imagined, and even though she had lost her golden voice, she remained an approachable and charming lady.

Other memorable graduations included the ceremony at CUHK, when as an honorary graduate I was able to speak on behalf of all the honorary inductees. The ceremony at the University of Glasgow also was very special, because at the event I found out that the person who had nominated me, Prof. John Lamb, had passed away only a few months earlier. The University of Padova holds its graduation in a 900-year-old hall. It was an especially rare honor to receive a degree from the second-oldest university in the world. Altogether, I have been granted nine honorary doctorates, and I earned one.

Now that I have seen the behind-the-scenes organization of how these awards are typically administered, I am aware of the difficulties involved in selecting a winner either when the number of contenders is large, or when you are scrambling to find a suitable candidate. Someone has to put forward the names of candidates. This requires a combination of word-of-mouth from CEOs, leading institutions, research labs, universities, and prominent citizens from around the world. Scientists and researchers do their work away from the public eye; breakthroughs and creative works are only exposed when they are published in scientific journals. Politics and jealousy play a part, too. After all, scientists are humans with the ordinary range of virtues and vices.

Without the support of colleagues and senior management, my work, no matter how pioneering, would have never been recognized. They were the ones who continued to put my name forward as a suitable candidate, and without them my feet would not be planted so firmly on the ladder that has led to my international recognition as the so-called "Father of Fiber Optics."

THE KAO FAMILY IN PHILADELPHIA AT THE AWARDS CEREMONY
FOR THE STUART BALLANTINE MEDAL, FRANKLIN INSTITUTE, 1977

Family Life and Travels

A life surrounded by a caring family that spans generations, even with the typical disagreements, is a blessing. Just as one needs a mentor in one's professional life to help guide and advise up the career ladder, so do we all need a wiser and more experienced person to guide us in our youth towards adulthood. This support network has fallen by the wayside in many areas of the world as globalization has dispersed family members far and wide. New ways of communicating, thanks to advancements in technology, have allowed this network to be revived. Now one can talk with children, aunts and uncles, grandparents, and friends who are thousands of miles away.

In the early years of marriage I was still struggling to build up my savings, beginning with the lowly initial sum of twenty English pounds. My father, remembering my youthful thoughtless waste of money on the purchase of twenty volumes of the *Encyclopedia Britannica*, wrote to me encouraging me to be frugal. He likely would have been horrified had he known of my tight financial resources.

Older colleagues at work were helpful with their ideas for budget holiday trips. Once, Gwen and I went on a camping trip to the soggy rural countryside in a leaking tent that we had borrowed from someone. My faithful old vintage Austin Seven took us from London to as far as Wales and back. I became ill before the end of that vacation with tonsillitis, and we tried to stay at a youth hostel in the mountains. In those early days, the rules were that guests were to be single, young, and on foot. The manager at this refuge in the Welsh hills took pity on us and let the two of us stay the night in a caravan on the grounds. That was a wonderful change from sleeping on the hard, damp earth, to hear the gentle trickle of rain on a solid roof instead of the steady beat on a canvas tarp of a tiny tent before finally leaking through the holes.

After the children arrived we became a family of four. The next five years were tough and fraught with anxieties. Many young families have the luxury of being part of a supportive extended family. My parents were halfway around the world. My in-laws, though physically very near in comparison, might as well have been as distant. My mother-in-law's tendency was to destroy all self-confidence and turn my wife into a stressed-out child. For our entry into parenthood we relied on books, especially Dr. Spock's guide, which was the top-selling primer for new parents.

Our new home in Harlow was damp and water dripped down the common walls between our neighbor's home and ours during the first winter after our son was born. The typically wet summers of England had not generated enough heat to dry off the plastered walls.

My son had terrible skin problems. At a few months old, he erupted with eczema and his whole scalp was covered in yellow scabs. This began after the visiting nurse suggested we dab milk of magnesia on small spots on his face. I shall never know whether this was the cause, whether he was overheated, or whether he had a milk allergy to the powder on which he was being weaned. The local doctor was not helpful as the scabs began to form, first on his face and traveling on to cover his whole head like a yellow, oozing helmet. The doctor suggested putting hot oil on the scabs and picking them off.

When this did not improve the condition, he scolded Gwen, complaining that she must not be following his advice diligently enough. The baby skin was so young and tender, no amount of soaking with oil made the scabs loosen their hold without tearing off the skin. It was too heart-rending for us to put our new baby through such torture.

Fortunately, after three weeks of this pain, Gwen took matters into her own hands and took her son to the hospital to see a specialist: "This is a common enough ailment in young babies. It may recur again when he is forty!" said this doctor. "Here, this zinc ointment will remove the scabs in twenty-four hours. Come and see me again next week."

Sure enough, the scabs simply fell off, exposing the very tender young skin underneath. We looked up the ailment in a common pediatric medical dictionary and there we found the specific treatment printed in black and white, just as the specialist had informed us. The local doctor was furious that we had gone to see the specialist without going through his office.

This same doctor tried to sabotage the private hospital room we had arranged for the birth of our second child at the same local hospital where our first child was born. The doctor insisted that a second child had to be born in the home. According to him, hospital stays were reserved for first-time mothers, as hospital beds were in short supply. However, we found that private rooms were available for a price, besides what was offered under the free National Health Service scheme of the day. Without much ado, we secured a bed. We later learned from a nurse at the hospital of the doctor's attempts to cancel our room. No one else would have been using the private bed if we had not reserved and paid for it out of our own pocket. Why was the doctor being so obstructive? Although the government system made it difficult to change our medical registration, we were able to find a sympathetic doctor willing to accept us as patients. He told us that he would handle the other doctor for us and smooth over the matter of the transfer of our medical files.

Though we had repaired fences with my mother-in-law, Gwen was still mentally traumatized by the pains of childhood. Her father had died when she was just twelve years old, and her mother had been left to bring up four children on her own, the youngest being only eight and the eldest, a son, barely eighteen. Alone in an alien land, without any knowledge of either the native language or culture, she was entirely isolated. With no professional skills, she had to frugally ration out the remaining finances. It was a terrible nightmare for her and soon she became a prisoner of the demons in her head.

Gwen tried to explain the situation to me: "My mother could fly into inexplicable and terrible rages without warning, yelling and cursing at us all. All our past sins would be remembered, plus ones of which I was not guilty. Those who were not yet at home could feel the frigid atmosphere immediately, even as they turned the key at the front door to enter.

The cruelest accusation fell on Gwen's young head; that she was the one responsible for her father's premature death and all the hardships that followed. This was a heavy burden for a twelve-year-old child, especially when constantly reminded of this with slaps to the head. It was a burden that haunted her dreams for decades after, even after I married her.

Excited at the prospect of two weeks of school camp in the countryside, she had waited and waited for her beloved father to return from his daily predawn trip to the wholesale fish markets in the city. The teachers had warned her that latecomers would be left behind. The school buses would not wait for them. She wavered between the two choices; she wanted to say goodbye, as it was the first time she had left home by herself and she wanted to give her father a hug. Yet if she left any later she might miss the bus.

The deadline loomed and she felt that saying goodbye would be useless if she was too late to go. Despite her mother's attempts to restrain her, she left running, clasping her small suitcase to her chest. Her short legs racing, the one-mile distance to the school seemed like ten. She arrived breathless and found most of the buses had already left. There was one remaining bus with its doors open. Less than a week later, her father suddenly died in the middle of the night from thrombosis—a blood clot somewhere in his brain.

Gwen loved school. From the age of four, when she had enviously watched her older siblings leave for school every morning, she had badgered her parents to enroll her, too. The school accepted her in the fall of 1939, before her fifth birthday. A few weeks later WWII broke out. Her mother even blamed her for the war.

My first holiday with my small children was to a holiday camp in the off-season. Now part of history, these camps used to be in seaside towns in the UK and consisted of small single-room cabins with limited cooking facilities. There were red-coated staff members to organize merry sing-alongs and games for young campers. All of the happy campers were awakened with music and encouraged to join the fun. It was a sort of forerunner to the Club Med idea for families. In the off-season many of the programs were not in operation, and we were happy not to have to endure the forced entertainment. The most memorable event occurred on the second day when Gwen strained her back. She was in agony and bent over for the rest of the week. The doctor suggested bed rest, but bed rest was intermittent at best as childcare is not really conducive to such an order. Gwen did not consider her own mother a fit person to take care of her children, so we could only look to ourselves. I became a model husband. My cousin, Vivian, who returned to Hong Kong after a year of study in the UK said on one of her last visits to me, "Your mother is never going to believe it! You didn't know how to lift a finger when you lived at home, and here you are cooking a fantastic meal. Even doing the washing and ironing and looking after a small baby!"

I found it very challenging to keep the house tidy and clean, to shop for food, and plan the meals. Gwen, who'd had to clean house from childhood, hated housework. To her it was so repetitious, boring, and never-ending. She was terribly untidy, while I liked everything neat and in its place. She complained that the house looked unlived in if I kept cleaning up after her. After a time of adjustment, Gwen became neater and I became less meticulous.

It took a number of years and conflicting diagnoses before the back problem finally cleared up. One doctor suggested complete bed rest for a couple months or more and another advised exercise. Her mother accused her of lying and faking it, an excuse to avoid the Sunday visit to her. Gwen learned to be cautious in what jobs she tackled and how she did them. Every now and then she would over-estimate her limits and I would come home to find her in tears and flat on her back.

In those years I became a handyman, too, repairing and mending things around the house. In England people are great do-it-yourselfers. After WWII, with short supplies of practically everything, "Make do and mend" was the prevailing motto. This habit has been passed on to the members of the DIY brigade. Hiring people to do repairs was expensive and took time.

I learned how to hang wallpaper and to do all the repairs that a homeowner finds necessary in order to maintain the value of such a huge investment. If I did not know how to attempt a job, I went to the library and consulted books. The skills I acquired in those early years were of great value as I progressed from ownership of this first house to the many later ones I bought in my life.

The job of which I am still most proud of was the laying of an intricate path around the edge of our garden in the last home I owned in Harlow. This involved digging up a path full of weeds, smoothing the earth, and putting down a layer of sand mixed with gravel. This was to be the self-leveling mix. I had ordered broken paving stone from the local council and this had been delivered and dumped on my front driveway. Laying out the path was like making a laborious jigsaw puzzle. My young son tried to help with his own small trowel and spade. It took me many summer evenings, working into the twilight, to complete the job.

I also learned to build a brick wall, though the completed effort was not extremely professional. The finished six-foot-high wall had a bulge and a curve. It was meant to provide extra privacy from our neighbor in our first house. My finances had been improving and I finally had enough saved for the down payment on a home of our own. It was a semi-detached, two-floor structure.

The house faced a barn with stables for horses. Although the road was lined with residential homes, this rural anomaly dated from a time when the area was farmland. The horse riders going in and out were a great attraction for my young son and he was often at the front windows watching for them. There was a fair amount of traffic coming through the road and one day Simon was witness to the traumatic sight of a car accident. He and his mother were admiring a dog, a Labrador with a beautiful, black, shiny coat as it bounded along with its owner. The dog suddenly darted into the road as a car sped around the curve. In front of their eyes the car hit the dog and the next moment it was lying on the road motionless.

In the summer of 1966, a young man, Philip, joined the family. Sent to attend boarding school in the UK by his parents in Hong Kong, I agreed to be his guardian and have him stay with us during his school vacations. He arrived as we were about to set off with our small children to go on a holiday to Europe. I was going to an ITT meeting in Brussels, Belgium. This was where ITT had its European headquarters. I planned to drive there in my Austin minivan, my first brand new car. It was an adventure for the whole family.

So Philip, a gangly teenager, dropped off his many pieces of luggage in the house and excitedly got into the car. We were to drive to Tilbury for the car ferry and to sail across the North Sea to Ostend. On the way, of course, we got a flat. We almost missed the ferry.

Leaving the family to explore Brussels with an extra person to watch the kids, I attended the business meeting first. Then free to be on vacation at last, we stopped en route to Paris to enjoy the miniature toy town of Madurodum, near The Hague. The children were fascinated by the replica houses that were as tall as they were. They were mesmerized by the working small trains going around and around, and enthralled by the tiny, busy cranes going up and down at a replica of the Rotterdam docks. They did not want to leave. Decades later I returned with Gwen to revisit this place. We could not recapture the joys of that day, and without the squeals of our excited children it was a somewhat dull visit.

The next day it poured rain all day. I had planned to make a stop in Amsterdam where we would stay a night so as to arrive in Paris in daylight. With the weather so gray and dreary I abandoned this idea and decided to push on south. A colleague from the STL Labs had a small apartment in Paris and he had arranged for us to stay there for a couple of days. Arriving now in the dark it took me some time to locate his place. We had to wake up the concierge to obtain the key. She was not a happy person.

As this was our first visit to the French capital, we excitedly explored all the well-known tourist sights. Driving on the right side of the road was a little frightening. The busy daytime traffic was heavy and continuous around the Arc de Triomphe. I drove around the circle four or five times before I managed to place the car into the correct lane to exit. While I fretted, the family admired the Arch from every angle. I dared not peek as my eyes were glued to the task at hand.

We climbed up the Flèche of Notre Dame Cathedral, rode the lift to the top of the Eiffel Tower, and climbed the many steps up the hill to the Sacré-Cœur Basilica. The children thought it all fun, even visiting the Louvre. Then we drove to Calais to once again cross the English Channel by ferry, landing at Dover. The famous white cliffs of Dover were a memorable sight as I watched the coastline of England approaching from the decks of the ferry.

As we drove off the boat in Dover, I cautioned Philip to stay in the vehicle as he was wearing his flashy Omega watch, an item that the customs officer would pounce upon. It was likely, with small sleeping children in the back of the van, that the officer would give us a cursory glance and wave us through immigration.

Philip's parents had provided him with all the toys of a teenager, including the latest Leica camera. They were highly taxed and expensive items in Britain at the time. Europe in the 1960s was still austere and not entirely recovered from the ravages of WWII. Meanwhile Hong Kong was a duty-free shoppers' paradise.

Being a curious young man and needing to stretch his legs, Philip casually sauntered around the car. The sharp eyes of the customs officer quickly spotted his glittering gold watch. We found ourselves paying for Philip's folly. For Philip, this might have planted the seed that led to his adventurous travels later in life. He was at an impressionable age. However, such an early introduction to international travel for my children and the consequent numerous around-the-world journeys we made as a family, had quite the opposite effect on them. As adults they have little appetite to explore the world.

Still intent on camping, I purchased a large, new tent with a sewn-in ground sheet. This was a luxury tent in comparison to my first borrowed one. I could stand up inside this one. Now the owner of a four-door sedan car that had a rooftop luggage rack, I was able to carry much more gear to add to the comforts of camping. The family set off the following year for a second holiday in Europe. I planned to go as far as Switzerland and to see the Alps.

I drove across as much of France as I could in one hop, entering Switzerland somewhere near Basel. My memory of that town is somewhat dimmed by the passage of the years, though I remember that there were many colorful flags hung out and the buildings had a medieval, sandy appearance. My first view of snow-capped high mountains was breathtaking. In the summer weather, the sky was bright blue and the air sweet smelling of hay and grass.

Our luxury tent was nothing compared to those owned by the many German and Italian campers. Theirs came with two rooms with a front zippered door, together with sun awnings and window flaps. Small carpets were laid before the front entrances and they sat on garden folding chairs in comfort to enjoy their meals cooked on fancy camping gas stoves. Even the campsites were luxurious. Hot and cold water shower facilities with sinks and nice toilets were provided. Water pipes were fixed so that campers didn't need to walk far to satisfy their needs. It was all very orderly, unlike those I had experienced in my earlier camping days in England. There the facilities were sparse and unpleasant. The sites were commonly on a farmer's field and campers barely tolerated. In the rest of Europe it seemed to be big business and campers set up tents for a long vacation in the same spot with comfortable cots.

We were definitely the poorer neighbors, resting the night on the hard ground in sleeping bags, cooking on a spirit primus burner, and sitting on the ground to eat our meals off a low folding table. We did not mind, as it was one of the best experiences in our more youthful days.

The mountains were awe inspiring, especially from the top of Mt. Juneau. One day we took a trip up towards the mountain peak in a cable car. The peak was shrouded in fog and the cable did not run to the very top that day. Another day we visited the caverns of stalagmites deep underground. Driving in the Alps, I wondered why the other drivers drove so slowly as I overtook them. I soon learned why.

The mountain roads zigzagged along the steep sides and hairpin bends were frequent. I suddenly found my brakes were not working well.

The linings, with my constant reliance on the brakes to slow the car in fourth gear, had been worn down. The wiser and more experienced drivers drove in second gear, using the engine to slow the car down without using the brakes. I did not dare to drive any more down the mountainside. My imagination envisioned the car going faster and faster only to end up smashed over a precipice. Parking the car in a rest area lay-by, I walked to the nearest house and got help to find a car mechanic. A young Swiss man soon appeared: "No problem. You must pump the brakes and drive. Then in the valley below, you can get it repaired at a garage."

But how could I drive with no brakes? The young man offered to drive the car down and apprehensively we sat in the car as he did so, at quite a speed, too.

The repairs would take a couple of days in Interlaken, so we found a bed and breakfast in the village. It was very cozy, the room had a huge bed with thick lace-covered down quilts. A railway line ran to the next village. It was a nice interlude in our travels.

From Switzerland I drove over the Alps via the Simplon Pass into Italy. We saw our first glacier and stopped the car so that we could all experience the novelty of standing on glacial ice. Through Italy we arrived in Venice at daybreak. It turned out to be the best time of the day to visit this crowded tourist destination. There were few people around at seven in the morning. The Doges Palace had a fantastic blue hue that I captured in my photographs. It looked like a wedding cake. My subsequent visits to Venice were amidst crowds in the heat of summer and I prize this early, unrepeatable impression. And then it rained and rained.

Camping in the rain with small children is no fun. We longed for warm dry weather. Instead of exploring Venice more, I turned westwards to Genoa in the hopes of finding warmth, sea, and sand. That is where we ended our vacation, camped near a sandy beach, where my daughter befriended an Italian child and together happily built sand castles. It was time to drive home again.

My parents did not meet their daughter-in-law until the year my son was about to start his primary schooling in Harlow. That year I made my first trip back to my parent's home in Hong Kong, together with my wife and two children. I had left Hong Kong by boat to sail to England over ten years ago. The Hong Kong I now came back to was much changed.

The population in Hong Kong had swelled with the addition of hundreds of thousands of refugees from the Mainland escaping poverty and politics. Flimsy squatter huts built of discarded sheet metal, cardboard, and wood lined the steep hillsides behind Victoria Park in Causeway Bay. A year or so later, after my home visit, terrible fires raged down these hillsides and elsewhere in the colony where squatters had settled in great numbers. Much was lost as lives, homes, and possessions fell to the flames. It took these tragedies to finally move the uncaring colonial government to take action. This was the beginning of a public housing policy. Those first seven-storied buildings were stark, providing just one small bare room for each family with a shared bathroom facility at one end of each floor. It was only marginally less wretched than the destroyed shacks on the hillsides. The estates were prone to crime and rape. It took the example of the newly independent Singapore, which built much better quality, modern apartments, to prompt the colonial masters to improve their standards.

Some of those original buildings still stand today, now converted into workshops and factories for small businesses. With a bit of imagination, you can envision the living conditions of those earlier times when these sites were homes to families. Fine new public housing is now going up all the time to give homes to the burgeoning communities. Whole new towns, in what were rural areas, have sprouted up since the late 1970s, like so many long needles pointing skywards in fields of concrete. People lead busy, industrious lives. Now there is a huge well-educated middle class instead of the previous imbalance where the poor were the majority, with the wealthy few ruling over them with little concern for humanity.

My parents had moved to North Point where a large number of people originally from Shanghai lived. Their apartment was on the first floor, a short walk from the busy Kings Road.

On this first return visit to Hong Kong, I found their bathtub had been used to store their books and had never been used for its intended purpose. Arrangements were made to take the kids to a clinic of a family friend, a Dr. Su, after it was closed for the day to use its facilities for their daily bath. It was a trying visit for everyone. There were uncomfortable, hot sleeping conditions as air conditioners were still a rare and expensive item in households then, an unfamiliar daughter-in-law who did not know the mode and manners of Hong Kong culture, and fractious children who did not want to go to the playgroups to join small kids who babbled at them in a foreign tongue. We went to numerous dinners with older family friends and relatives for my parents to proudly show off the grandchildren, and to welcome a new daughter-in-law and the return of a prodigal son.

It was with some trepidation that I left my family while I went off to Japan for a week to network. It was a stressful time and I was very glad to get the family back to our own home again after a three-week absence. As an earnest engineer working passionately on amazing projects, I had an inkling of the enormous impact it might make on the industry. It was always hard to tear myself away from the labs to go home in the evenings. The ideas for how to make the glass more transparent and the methods for experimentation, plus how to solve the intricate mathematical equations to prove the theories swirling in my mind all through my waking hours, and possibly in my dreams. It was not unusual to find that I had new ideas in the middle of the night, just before I was fully asleep. I would always want to tell them to my dozing wife. "Why don't you write it down before you forget it in the morning?" she would groan and roll back to sleep.

As Gwen had a basic understanding of the physics, having studied the subject in her university days, she was the sounding board for all my ideas. Talking to her helped me clarify my thoughts and explain them in a way that was less technical. Or she would point out a knotty problem that had not yet crossed my mind. At work in the same labs on a half-day basis, as she was in the computer section working on software programming, she had the responsibility of designing the programs that solved the many mathematical equations my team and I passed on to the group there. In those early days, FORTRAN was the computer language used for scientific pursuits. She wrestled with the Bessel functions that she said she could never understand in her earlier mathematics courses at university.

However, Gwen thought out how to set up the routines to get the computer to reiterate calculations over and over until it found a near solution to my equations. Many of the routines were not standardized then and had to be written out from scratch. So she often came home with ideas swirling in her head, too. Of course there were also the bugs in her programs that had to be found.

Before the labs bought the first model of the big IBM computers, we had been using a valve computer. For this computer the programming language was Assembler, which was similar to machine language. The instructions, instead of using punch cards like the IBM computer, were put on punch tape. One held the tape up to the light to see the holes punched out that corresponded with an instruction. It was easy to mistype. Gwen would bring home coils of paper tapes trying to find the typed errors that were causing her programs not to work correctly.

The children played with the discarded tapes. Sometimes their mother would take one child with her to the labs in the evening. Having found an error she was anxious to re-run the program in the hopes it was the only error. There were lots of tapes there for the kids to rummage through and unearth from the bins. Small children love to make messes and tear up paper. It was surely this early introduction to computers and the technical world that led both of my children into that field of study.

Later, with an IBM computer, it was pages of FORTRAN printouts to be scanned for errors. Gwen used to tear out her hair with frustration in her attempts to find the elusive bugs. The printouts became scratch paper for the kids to draw on.

In this way we were quite a husband-wife team. We shared the load of childcare and housework. When I came home late, by which time the children would be getting tired and cranky, I was often in the doghouse. By 6:30 in the evening, an hour or more after the normal time for my return home, she would be expectantly walking the children up and down the walkway outside the home. On a summer day in the UK, the sky would still be very bright and sunset took place as late as 9 o'clock. The lab was only ten or fifteen minutes away by car. Mrs. Smith always left by lunchtime, leaving Gwen's hands full to entertain the children and to do mundane household chores in the afternoon.

"Look for Daddy's car coming from over there. It will be the next one for sure. Let's see who can be the first to see him," she cajoled the impatient toddlers. Sometimes she was lucky, but mostly I barely arrived home before the children were put to bed: "The project is so enthralling that it's hard to break away to come home. One day it will be a fantastic product. Just think, I will be famous then." I tried to pacify her with my tale of future fame and fortune. I myself only half-believed it, though I had full confidence that my ideas would work.

In 1969, I joined an ITT business trip to Russia. The Russian trip was a PR exercise to sell the business of ITT. In that Cold War era, I searched my room for listening devices. Though nothing was ever found, my colleagues and I all felt the need to be cautious. So it was alarming when the limo driver taking us to a destination was pulled over by the police. There was a shouting match that we could not understand. But we thought it was better to disappear fast ourselves and the four of us got out of the car and walked away quickly, merging into the crowds.

To be part of an official visiting party anywhere is to be treated like kings. Our hosts made sure we saw the national monuments and sites of historical significance. Without standing in the long lines, we were whisked straight in to view Lenin's preserved body. Normally hard-to-procure tickets were found for us to watch the Bolshoy ballet at the Opera House. We viewed the treasures in the Kremlin. I wandered by myself into Gum, the big department store in town. As the doors opened in the morning I found myself swept in by the huge crowds of shoppers. The shelves were mainly bare and customers were massed densely in front of the few counters that had merchandise. Successful purchasers were given a sales slip. They then had to line up to pay for it at another counter before returning to line up again to collect the goods. It seemed to be a time-consuming task to buy anything.

The business side of the trip was spent visiting the Post and Telecommunications network industries. We hoped to observe and learn how ITT products could match up with their requirements. Suddenly there was an unexpected request from the Russian Office of Telecommunications for me to give a talk on circular wave-guides to one of their top brass. I was introduced as the up-and-coming expert on circular wave-guides.

On an early ITT trip to China, when it was just opening its doors a little wider, our group was entertained to many meals and toasts with fiery alcoholic beverages, especially the much-touted Mao Dai spirits fermented from rice. Subsequently I wised up and learned to say I was a teetotaler at such banquets.

These business trips to promote the company might sound like free vacations, but in fact they were very tiring. We were up early because the hosts needed to drive us to whichever factory was on the schedule for the day. At the factory we listened to hours-long recitations of facts, figures, and glorious advances. We sat through long, polite speeches made both by our side and by the host, and then continued sitting through the exchange of gifts and photo sessions. We then walked through their production lines and made observations, exchanged technical facts, and perhaps another department might want to give their own take on the facts and figures. Then we would be entertained to a lunch that was full of toasts to this and that. While I found the food good, my Caucasian colleagues found it strange and new to their palates. Worried about digestive problems, they were wary of eating the food, yet they could not offend their hosts by refusing to eat. We mostly fell into bed at nights exhausted by the day's excursions and meetings.

The world was beginning to open up and engineers were rapidly becoming better traveled. I was encouraged to participate in conferences held in other countries, to spread nonproprietary knowledge, and to be aware of where the frontiers of research were in my field. I was busy exporting my ideas of the possibilities of fiber optics for telecommunications far and wide. This was the way to get research labs excited and for them to find money to further their work. I held to the theory that the more people who were engaged in this field, the more people would see the light.

There was a phrase for those engineers who had gone on a trip from England to the United States for STL on business—"been to." It was a title looked on with envy by colleagues who had not yet gone. It denoted a rite of passage in one's life. Many of the engineers, including myself, had visited the European ITT headquarters in Brussels or the facility in Paris. A longer flight on a plane was still a novelty. My first trip to the United States in 1967 was to Bell Labs in New Jersey. It was an extended stay as I went up north to visit the Corning plant in Ithaca, New York, as well.

I felt overwhelmed by New York City. Walking along the avenues with my head turned skywards in awe of the skyscrapers, I suffered from the normal neck cricks of first-time visitors. One night was spent in the YMCA that was located in a rundown-looking area on the city's west side. I felt unsafe and nervous. Perhaps I had seen too many of the American-produced gangster films. Later on I was to visit the U.S. so often that I finally lost count, and as I moved up in the company so did the star value of my hotel choices.

Being a resident of a country is a different feeling from being a visitor there on a relatively short business trip. During a short stay everything is a new experience—the currency, the food, and how to pay for it. Every culture has its own little ways of doing things. The natives are generally friendly and then one returns home, after which the trip is largely forgotten in the days of work and family life.

After I moved to Virginia, I could better appreciate the problem Americans had with racial divides. As an Asian family in the provinces, we were stared at and occasionally verbally attacked. My children were bewildered by this phenomenon. I had never been exposed to this behavior in my life, but to Gwen, who had grown up in Britain, it was all too familiar. She taught the children to turn the other cheek, to smile with pity on the ignorance being shown, and to take the opportunity to educate the unschooled minds.

By then we were a well-traveled family. During my four years of teaching at the university in Hong Kong, I was on expatriate terms as I had been hired from the UK. This meant that I was entitled to an annual paid leave to return to our home in the UK every summer. This, besides allowing me the opportunity to recharge my technical skills and to work with my ex-colleagues in research, gave me an ideal way to return by roundabout routes to London and then back to Hong Kong.

We fit in visits to Austria to see our former au pair, who lived in Vienna. We still keep in touch with her after all these years. It is marvelous to know friends in foreign cities. The usual sights were always taken in, of course, like the beautiful Schönbrunn Palace, filled with Hapsburg treasures, the view of the city from the Vienna Woods, St. Stephan's Church, and the Opera House. Our old au pair even took us out to visit her home village to show us her mother's house with its outdoor toilet. Unlike the cobweb-filled dingy outdoor toilets of the Victorian childhood home Gwen remembered in London, this one was spick and span with a seat of highly polished wood. Still it must have been as unnerving for a small child to use it in the dark of night, holding a wavering candle or a dim flashlight. Gwen said she used to sing out loudly to announce she was coming, all the way down the dark steps and into the black, black night outside. The ghosts, spirits, and bogeymen would be scared away if they heard her coming—an early warning signal for them to run out of her path.

Then there was the memorable year that we flew via Greece and walked up to see the Acropolis in Athens. Today this antiquity is roped off and one can only admire it from afar. Back then we could walk along the pathways where historic feet had trod and be in awe of the past centuries. We had had to seek help from the local tourist advisory group at the railway station for accommodations. As it was sponsored by the State, we expected it to be an honest business. We were very much into arranging our own itineraries and had a dislike of being shepherded around in a group. However, the first hotel we stayed at turned out to be the typical tourist trap. At dinner that night we observed generous helpings of lamb stew going to guests at a nearby table. We were served tiny helpings, especially the children, who were the most hungry. Grudgingly the waiter gave us each a little more. The next day we dropped by a travel agency in another street closer to the center of the city: "That hotel is a rip-off. I can book you into a far nicer hotel, and with air conditioning, too, which is good in this hot weather—and for a lower price than that one!"

I can still remember the name of the nicer hotel, Hotel Cristobal. From Athens we booked a day trip to one of the famous Greek islands. We were quite adventurous in roaming on our own in strange cities and getting thoroughly lost. Though we were always lucky and managed to land on our feet with help from kind strangers.

There must have been an air of innocence about us or perhaps we exuded a transparent honesty. In 1970, at the beginning of my four years as an academic in Hong Kong, we were temporarily housed at Shatin Heights Hotel. We went one evening to a restaurant down the road in the village of Tai Wai and enjoyed a family dinner seated outside. To my dismay, I found I had left my wallet in the hotel and had no money on me at all. Leaving the family as hostage, I said I would go back to the hotel and return with my wallet. "No need, no need!" said the proprietor. "Just come back and pay me tomorrow."

At the end of my contract with the Electronics Department, we took a long route on our way to a new destination. I was to rejoin my old company, ITT, though it was to be at a different location in the United States. As there was a conference on fiber optics, one of the annual IOOC events, in Kyoto that year, I decided to route the journey from Hong Kong via Japan, San Francisco, Los Angeles, Houston, New York, and then on to Roanoke. I had the good fortune of being friends with a Japanese academic who had been part of the fiber optics research team at Southampton University in the UK. He and his wife entertained my family along with two other former colleagues of his from the university on sightseeing tours of the city. At the Golden Temple, we sat down to a Buddhist vegetarian meal. It was so delicately laid out and artistically arranged that we were reluctant to disturb the food.

This was not my first visit to Japan, but it was for Gwen and the children. Gwen fell in love with the ceramics of the country and in her many subsequent visits to Japan she always went straight to the ceramics floor of the Japanese department stores to find something to add to her supply of kitchenware at home.

Japan is the country that first took to the idea of fiber optics, and in those early days I made many friends among the researchers in the major Japanese telecommunications companies. The visits to Japan are too numerous to remember except for the highlights. I learned much about its culture. Never wear your ordinary house kimono the wrong way around. The front should be wrapped and tied left over right, even for women. The other wrap, right over left, is only for mourning. Gwen was shown this with much sign language at a small Japanese inn in the mountains. In Japan we learned to exchange our house slippers outside the bathrooms for slippers meant especially for use in the bathrooms. In Japan we also enjoyed soaking naked in the steaming mineral waters of the open air hot springs, with the men separated from the women by thin screens. We also learned to sleep on tatami mats on soft rolls of cotton bedding.

From the stop in Kyoto, where I met many old colleagues at the conference from my previous industrial life, we flew on to San Francisco to stay for a day or two, which was just sufficient to say that we had seen the Golden Gate Bridge and ridden the trolley cars to the wharf and back. In Los Angeles we visited Universal Studios and, of course, Disneyland. The lines for the rides were amazingly long and the children were very frustrated, as we did not have the time to stand around for hours. Looking back, I wish that I had made a more leisurely travel schedule. Why do the young rush about like time is about to run out? Why did I always count my pennies so carefully? How was I to know that this was not necessary as more was to come? Hindsight always comes too late.

If I had been wiser and worried less about my finances, I would have taken my uncle's advice and bought an apartment for investment purposes before I left Hong Kong. He would have managed it for me. In the same way, if I had bought an apartment in Los Angeles, as a good friend there had asked me to do during one visit, his wife, a real estate agent, would have managed it. I would be a millionaire many times over by now. My salary was already many times more than my meager initial paycheck and my savings had grown substantially from a paltry twenty English pounds sterling. Life now contained more comforts.

I stayed in Roanoke for eight years before moving to the more cosmopolitan areas of the East Coast. During those years, I lived a comfortable suburban life in Middle America. While my family learned to be part of the local community, I was immersed in my work at the plant. This involved a fair amount of traveling and visiting ITT headquarters in New York and other ITT plants in Raleigh, North Carolina and around the world. Gwen calculated that I was on the road 40 percent of the year.

In 1985, I was on the road for 100 percent of my time because this was the year when I voluntarily took up residence in Germany. My daughter was at her first job with AT&T in New Jersey and she was living away from home in her own apartment. She drove home, a two-hour drive, only on the occasional weekend to visit us. My son was working a short distance from our house, so he chose to live with us for a while after graduation. I was free to follow my dreams, especially as ITT had given me precisely that freedom when they appointed me executive scientist. I asked to join the SEL Lab in Stuttgart, Germany for one year.

In Germany shopping hours were strictly kept, even by the smallest of family-owned businesses. One morning, as Gwen drove me to work, we passed by a small shop. The owner was busy setting out his shelves of goods in front of his store. To all intents and purposes the shop was open for business. But it was still five minutes before the official shop hours. It would have taken her at least that amount of time to find the item she wanted to buy, but we could not persuade the shopkeeper to allow us in. No matter how we asked, the shopkeeper simply responded, "It is forbidden. It is not yet 8:30 a.m."

I am glad we did not live in Germany for long. A year or so after our time there, it went through a period of racial disharmony and I would have undoubtedly felt its effects. I have rarely met with racial discrimination. Gwen has been a victim more often, beginning with her experiences as a child in the UK. Her encounter in a park in Ditzingen, the small village we lived in outside Stuttgart, was a warning of the violence that was coming to the rest of Germany, when lighted petrol bombs would be thrown into immigrant homes.

Two men were digging trenches in the flowerbeds of the small park that Gwen crossed through on her way to the shops. One was an older man and the other a teenager. The teen yelled some German obscenities as Gwen passed by and purposely shoveled out clods of dirt at her feet. Puzzled, Gwen walked a few steps further as she analyzed the foreign words in her mind. Unhappy with the insults, she formed a suitable German phrase in her head and turned back to confront the young man: "Was fur eine Mutter haben sie? So schlecht ein son hat sie!"

It was terrible German, but she had made herself understood. The boy rose and made a threatening gesture, only to be restrained by the older man who pulled him back down.

Why do people feel threatened by those who look different? Why does ignorance bring out the worst in human nature? Sometimes it is not ignorance but brainwashing from parents who pass on their prejudices.

Another time, Gwen came home and recounted a different story. Traveling alone to the United States, she was shuffled into an overworked immigration hall at Kennedy Airport in New York, together with passengers from several jumbo jets all arriving at the same time. The lines were long, and everyone was tired and jet-lagged, but this was no excuse for the conversations that occurred. Well-dressed, well-educated English businessmen (a boss and his two underlings) stood behind Gwen in the long stalled line. Way up ahead near the counters there was an Indian family with crying babies and agitated men who were trying to learn the formalities of entering a foreign country. The businessmen made some terribly offensive remarks about them.

Gwen reported that she simply could not stand hearing their comments. "As I am standing right in front of you, and I cannot close my ears and am forced to listen to your racist remarks—you are at liberty to say such things in the privacy of your own homes, but this is a public place. I would appreciate it if you stopped your conversation!"

The boss pursed his lips and remained silent while his subordinates, red in the face, stuttered and attempted to apologize. There was dead silence for a long while and then they awkwardly attempted to make small talk amongst themselves.

I am proud of Gwen, and I am also glad her self-awareness took longer to bloom, as she might have moved on to another plateau beyond my reach. By the time I moved to Roanoke to become an engineer after my academic break, the field of fiber optics had grown at an exponential rate. My hopes finally were being realized, even though it had taken a decade. It was to be another decade before fiber optics would become a common product throughout the world, and a decade more for fiber to change the modes of communication.

Without realizing it, I had grown into the role of mentor and role model. Family life moved onwards. The children grew up and left home to lead their own lives. We moved again to the other side of the world and began another life under a different hat. I would learn to be more circumspect and conscious of the fact that I would soon be of the generation that was considered wiser in the ways of the world. I needed to be prepared to give advice, and I would soon find out that this was not a simple task. Diplomacy, tact, tolerance, understanding, and forbearance are all required tools for such a role.

OUTSIDE THE CUHK UNIVERSITY SCIENCE CENTRE (2008)

Sojourns at CUHK

Thirty years makes a difference, but some things never change. The mosquitoes are as large and ferocious as they were in the 1970s. I had already been bitten multiple times in just the ten minutes I was waiting for the minibus to take me up the hill from the railway station to Residence 1. Residence 1 used to be the staff quarters.

As I waited by the railway station, vainly batting at the swarming insects, I could see the many new buildings that had been constructed around the campus. The latest additions were just a short way up the slope on my left. Low, yellow buildings from the 1950s had once stood there, as had the private residence for the head of Chung Chi College. That was in the days when each of the three founding colleges of the university had special houses for their own president. The university was established in 1963, the brainchild of far-sighted educators. However, the three constituent colleges, combined into a single university, had found it difficult to give up their autonomy. In those early formative years, bitter wars were fought and jealousies abounded over territorial rights. Each college had persisted in maintaining their own libraries and registrations, and administrations had clashed during the centralization.

Today the campus glows with a new enthusiasm that is expressed by the many bright, new facilities. The windows sparkle in the new administration building along that stretch of road, catching the rays of the morning sunshine. The sloping road is unchanged. Residence 1, which had been home for two years to my family in the 1970s, is still standing, higher up—tiered on the mountainside. At that time it was a newly constructed housing complex of six floors with two spacious apartments per floor. This was the first staff residence to be completed and was very extravagant. Looking up and shading my eyes from the strong light, the building now looks very shabby. The outside walls are discolored and stained black from decades of acid rain.

In 1974 I thought I had left CUHK and Hong Kong for good. There were no ties to bring me back on a permanent basis. My parents had emigrated to join me in the UK in 1967, and I subsequently had to leave them there for two years to fend for themselves. Then the two years became four. Though I visited them annually when I returned to the UK to stay current with the progress of optical fiber research, these visits were much too short for my parents. However it simply would not have made sense to have them return to Hong Kong, as my time there was to be only temporary. It was better for my parents to stay in the UK, especially with all the moving I would be doing in the United States in the years to come. Opportunities came out of the blue without my seeking them out. I wonder how my life would have turned out if I had not taken advantage of these opportunities. I will never know. One of my old professors said to me long ago, "Look, son, luck does not just happen to you. Everyone makes their own luck."

It was a great feat that my father had managed to arrange the long trip with my mother from Hong Kong to our UK doorstep. While they were in their mid sixties, younger than I am now, they appeared much older as compared to myself at the same age. In their mid sixties, they had decided it was time to retire and live with me.

Unused to the excitement of children running around the house and disorientated by jet lag, my father succumbed to dizziness within the first week of their arrival. Unfortunately, he fell down on the carpeted floor and cracked his spine. When the accident happened, Gwen called her friend, Shirley, a senior nurse at the local hospital. Fortunately she was home and rushed over to assess the damage. Shirley was efficient, calm, and reassuring. "Get an ambulance," she instructed, "He may have broken something at his age!" My father was groaning in pain. Even a pinprick produced loud cries of pain from him.

My father was not a good patient. In the hospital he groaned all night long, disturbing all the other patients in the ward. Shirley related this back to us with much amusement. She said that she had to threaten him with dire punishment if he did not stop. My father began acting like royalty. He demanded attention and kept ringing the bell for nurses to attend to him. I was embarrassed for him, apologizing to the hospital staff when they asked me to keep him in line. They made a formal complaint to me to keep my father in order. He came home a few days later with a plaster casing around his torso. He was now our problem.

Over the next six weeks he recovered, after which mother wisely decided it would be better to find separate accommodations. She had been a silent observer of the added stress and tension from their presence. Unlike what they were used to in China, we had no maids to help out. Mrs. Smith only came for a few hours each day to supervise the children.

When Amanda was six months old, well before my parents arrived, Gwen returned to work at the same research laboratory that I was working at. She was lucky to be hired on a part-time basis, leaving her free in the afternoons to take care of the children. Gwen was hired to write the software programs to solve the complex mathematical equations produced by the engineers. However, we first had to find a competent home-helper to take care of the children in the mornings.

How we were able to find reliable help is a story in itself. At first we applied to agencies dealing with au pairs from the other countries in Europe. The au pairs were young female students from good families who wanted an opportunity to live in England with a family for a year to improve their English. They would do light house cleaning, provide some childcare, and be part of family life. The family would provide them a small stipend, pay their travel expenses, and allow them time to take classes.

Our first au pair was from France. We should have known better, as she was an only child. She didn't even know how to boil water! We had to spend our time taking care of her. In addition to this, she had a boyfriend who was in the French Navy.

In the brief three months that she lived with us, Françoise managed to lose our three-year-old son while shopping. With Amanda in a stroller, it was a pleasant walk for the three of them to go downtown. However, Simon was a handful and so one time she told him to wait for her as she continued browsing with Amanda in the store. Alarmed at feeling abandoned, Simon burst into tears at being left alone, until some considerate shoppers took him to a nearby police station.

Fortunately Simon was safe. Françoise was horribly sorry, now understanding that she should have been more responsible, promising that it would never happen again. We were leery.

Then her boyfriend came to town. He asked to stay the night on the couch, as it was too late to get back to his unit. I reluctantly agreed and they both disappeared up into her bedroom after dinner. After midnight, Gwen prodded me awake. She had stayed awake listening for the young man to go downstairs to his couch. "We don't know what they are up to in the room! How do we know whether her parents would condone it? We are responsible for her welfare, after all."

I thought it was Gwen's job to sort it out and left her to do whatever she thought was right. After thinking it over, she got up, knocked on their door, opened it enough to see that they were both asleep in bed together—and said loudly, "I think you should go down and sleep on the couch!"

The next morning, we told Françoise that she was being terminated and that she should leave with her boyfriend. This was the last straw for us.

Now knowing better, we gave the agency another try. We would be pickier as to the characteristics we were looking for in a replacement. That was how Trudi joined our household. Being from Austria, she was a more sensible girl, though she said Simon was a crybaby and not a boy! She stayed with us for eighteen months and in that time we got to know her very well. She had studied the works of Goethe and Bertrand Russell, and I enjoyed our discussions on philosophical matters. Trudi became so much a part of our family that she was like a daughter. Even when she returned home to Austria, married but sadly without children, we kept up a correspondence and have met up again several times in the thirty years since.

Au pairs stay only for two years at the most. Changing help at such frequent intervals, just as a routine was settling in, was not a very good arrangement. I needed to find a more permanent solution. We advertised locally in the newspapers. Amongst the many applicants was Mrs. Smith. We picked her letter from the rest as sounding the most promising. A woman in her forties, she was warm and cheerful, with grandchildren around the same ages as our own children.

So when my parents came to live with us, Mrs. Smith was already a much-appreciated fixture. She was the solution to helping my parents learn to live on their own. When I left to take up the post to join the Electronics Department in CUHK in 1970, Mrs. Smith agreed to take care of my parents three mornings per week. She kept me informed of their health, their living conditions, and their diet. Shirley was also a great help, as she continued to watch over them by dropping in to visit.

In 1971, when we moved into the brand new campus at CUHK, my young daughter stood every day near the lower exit gates, which at that time was the only exit from the campus. The school bus waited just outside the front door of the Chung Chi College president's home. Every child living on campus relied on that bus to get them to and from their schools in the city—unless their parents had arranged alternative means of transport. Every morning Amanda and her mother would walk down the hill from Residence 1 at the early hour of 7 a.m. to join a line of children ranging from small kids to gangly teenagers standing with their heavily laden school bags, waiting for the bus to appear. There was always an adult to supervise the children. This was necessary due to the mixed cultural backgrounds of the campus families. The children from the Western worlds could be relied on to be hyperactive and boisterous. The children from the Asian worlds would stand in timid awe, wholly at the mercy of what was perceived to be improper behavior. The weather was too hot already, even early in the morning.

"You can go straight back to the end of the line," Gwen yelled one morning as she supervised the children. She didn't care whose child she reprimanded. In those days she was using her power for the benefit of the downtrodden masses of colonial Hong Kong. Injustices were a sore point for her. "He's the president's son!" a child whispered. "He always gets on first."

The school bus route was torturous. The schools were scattered all around Kowloon. The route was mapped out as each school had different school hours. It was quite a nightmare for everyone concerned. Unless the school was government run, they were all independent. Families who could afford to pay for private education for their children avoided the government schools, which were perceived to have low standards and substandard facilities. Private schools did not try to have the same school hours as other schools, as they were answerable only to their own autonomous school boards.

Amanda was the last student on the bus when going to school and the first one picked up when returning home. The trip took an hour or more one way. The bus was not air-conditioned and she was hot, sticky, and cranky by the time she got back home. When she was nine years old, I felt more comfortable letting her ride the train into town with a couple of older students attending the same school, but it was still a hassle. The train from the university only ran once per hour as it was on a single track.

I told my daughter not to ride on the steps, but to push her way into the car. Most of the time all she could do was to squeeze on to the steps. We are very thankful she survived these adventures.

Searching further back in my memory as I board the campus minibus for the short trip up to my old home, I see myself as I was then—a young-looking academic who seemed barely old enough to be out of school, let alone the head of a department.

Taking a leave of absence from my company in 1970, I rented out our home in the UK and took up a two-year contract. I was unaware of the significant change my life would take from what I then considered to be a minor decision. I fully believed we would be returning to pick up our life where we had left it, in the small town of Harlow in Essex County. The two-year contract was to become a four-year contract and we never returned to the UK.

The small apartment in Harlow in the UK on the fourth floor of a six-storey building, where my parents made their home, had been located less than two miles from my house. The shops, library, and post office, as well as many other facilities, were all within walking distance for them. However, my mother suffered from agoraphobia and also carsickness, so she rarely left home.

When I left Hong Kong for the U.S. in 1974, my parent's lives were fairly well established. Father had made some contacts with the local Chinese community, occasionally traveling up to London to meet old friends. He would always then insist that they drive him back home. He occupied much of his time copying essays and poems. He said he was keeping an almanac in which all sorts of important facts were written. Piles of old magazines and newspapers were stored in the small apartment. Then mother was diagnosed with cancer in 1976. She lived for just six months in ever-increasing pain though she never complained. She ended her days in the same hospital where her husband had been treated for his cracked spine.

After my mother's passing, Mrs. Smith continued to look after my father. It had been his wife all along who had kept him in check. Now that she was no longer around, he became more obstinate in his desire to be as free as he wished. Mrs. Smith complained he would not allow her to dust or to touch any of the piles of old newspapers, magazines, or paper bags full of assorted items. He would not change his clothes or even bathe once a week. I would set up arrangements for a nurse to help him bathe, but as soon as I left he would cancel the service. The apartment was dirtier and dustier each time we visited. Mrs. Smith was powerless and was yelled at for daring to move anything. Gwen used her short visits to the UK to try to clean up as much as possible. She did not stand for any nonsense from my father.

In 1987, shortly after my unexpected appointment to be the third vice chancellor of CUHK, father was moved to Hong Kong to live with me. The move was not without its problems. We only had time to help him sort out his belongings in his home. He had to fly to Hong Kong with just one small suitcase.

"We can get all the newspapers you need in Hong Kong, and the magazines. Let's throw the old stuff away! The bags are of no use now because you will collect others later."

Gwen cajoled him with half-truths and tricks as she carted arm-fuls of trash down four flights of stairs to the large trash containers that served the occupants of the building. The elevators, as usual, were not working. It was backbreaking work and the trash container soon quickly filled to the brim. Unfortunately, I did not have the time to find homes for the beautiful Chinese gowns, embroidered with elegant designs and silk lined. As Mother was so tiny, her clothes, which father had left undis-turbed for a decade, fit no one, but the material would have been of value. Sadly, these also were consigned to the trash bins.

I sorted through the rest of the clothes and found that those of my father's were not wearable. He needed new clothes.

"These sets of underwear and two shirts will be all the clothes you need to put in the suitcase. We will buy everything else in Hong Kong. You will have room then for all the other bits and pieces you might like to take. The city council has said they can arrange for the furniture to be carted away for use elsewhere. And they are willing to clean up the rest. So you can just leave without having to do anything except to lock the door and give the keys to the caretaker."

I did not have much patience with his habit of hoarding each and every item that passed by his eyes. Having moved home so many times already in my life, I was used to clearing out and starting anew. It is a painful process, but a necessary and practical one.

Father flew into Hong Kong sometime before Christmas of 1987. I barely had time to settle into my new job. He came with his suitcase crammed with old pens and pencils, his precious almanacs and diaries, more old clothes than he needed, worn out shoes, and paraphernalia that my wife said was junk.

Mother had died in 1976, two years after I left CUHK. After her fu-neral, Grandpa had come to visit me in my home in Roanoke in the U.S. He stayed for several weeks and also went to stay a while with my younger brother in Washington, D.C. After sampling our lifestyle, he preferred to return to his own. In his seventies, he was a healthy and alert old man, though fixated in the ways of the past. He returned to the UK and continued to be an independent old man. His health had its ups and downs, but nothing of a serious nature. He was as healthy as a horse into his eighties. I think it was the lure of the reflected glory of my high office that enticed him to give up his independence. The following years of life with my father gradually developed into a need to firmly control an obstinate old man with deteriorating physical strength, who was firmly set in his ways.

Today I could have taken the quick fifteen-minute walk up the hill to the Residence 1. The medical clinic is temporarily housed there while its offices are being renovated. However, I am early for my appointment and in waiting for the minibus, I can sentimentally muse on the past.

I had left Hong Kong in 1953 as a callow youth of nineteen, half a lifetime ago. My continuing education and working in a foreign country had turned me into an Anglophile. I acquired many of the comfortable habits of my adopted country. From rented housing I had steadily improved my status and eventually become the partial owner of a large house with a garden in Harlow. With a wife and two small children, it was a satisfying life. My elderly parents had recently emigrated from their part of the world to join me in the UK a year or two earlier. I then had made the decision to leave them there on their own for a while. Still it was to be only for two years. In the end it became decades before my father came to live out his days with the family. My mother died waiting.

When I arrived with my family back in Hong Kong, laden with our bags, much of the campus was still barren rock and a busy building site. The classrooms were scattered into several temporary locations. The three colleges, which had been joined, still occupied their original locations— one on the Island and one in Kowloon. The third already was located on the very site that was to be turned into a campus for the new institution for tertiary education. Chung Chi College campus lay at the bottom of a hill. The undeveloped hill, covered with shrubs and wild vegetation, overlooked Tolo Harbour and the distant islands. The far shore of the bay was an undeveloped hillside with a handful of small fishing villages dotted around. The view was stunning. The railway had been in place for some years, so although this new campus would be considered remote to the urban areas of the day, there was a transport link already in place. The nascent Electronics Department was still considered part of the Science Faculty. It was temporarily housed in an old building on Caine Road on the Hong Kong Island side.

In the intervening years Hong Kong was no longer the city I had known of old. It, too, had moved on. Gwen, who had never lived in Asia, found it hard to adjust. My work absorbed me and lively colleagues surrounded me. It was not as easy a transition for my family. We all needed to learn Cantonese. To this day my Cantonese remains heavily accented. Everyone can tell I am a Shanghai native as soon as I open my mouth, but Gwen has done much better. From speaking the village dialect used in her home by her parents, she has progressed to a confident and capable conversationalist. Initially her accent raised eyebrows and looks of derogatory amusement, causing her much pain and a loss of confidence. However, Hong Kong is now full of provincial accents of all kinds and the locals are not as disdainful. They flock to Shenzhen in the tens of thousands daily and cross the border to mainland China for discount shopping. They haggle and bargain with shopkeepers who in turn find their city accents just as amusing.

Our accommodations for the first few weeks were in the now-defunct Cathay Hotel in Causeway Bay, next to the French Hospital. The lamps in the room were dim because the management exercised economy. It never occurred to me to change the light bulbs as we lived in the semi-darkness. Family life in one room in a hotel for three weeks was a penance.

My son entered Primary Four at my old alma mater, St. Joseph's, and my daughter enrolled at the Sacred Heart Convent on Robinson Road. We had tried to enroll them into the English Schools Federation but were brusquely told there were no vacancies and that the schools were for English-speakers only. We were anxious for them both to learn Cantonese anyway and had decided the local schools should be tried first. The children were really thrown into the deep end. Amanda, younger and easier going, soon made friends and was happy. Our son, older by two years, became moody and difficult. In hindsight we should have been more aware and helped him cope. Instead, under great stress ourselves, he was yelled at and misunderstood. He resented the learning of a new language and lost confidence in himself. Years later, in the process of moving house, I found old school books from that era. We were horrified to read one English essay, and to note the teacher's ominous comment in red: "Come and see me after class!"

I finally found an apartment on Robinson Road and thankfully moved out of our one-room home. The landlord insisted on a two-year lease. We thought he was a nice guy, taking us all out for a hearty dinner of snake soup in one of the famous back alleys, where the snake markets thrived. That was before we tried to break our lease to move into university campus quarters. Then he was not so nice. He wanted the rents fully paid up, even though new tenants were found. After much negotiation and assistance from the university, we were able to get the sum down to six months' rent as a punishment for breaking the lease.

That building in Robinson Road is no longer there. In its place is a multi-storied, luxury residential building with a marbled lobby and fancy-uniformed doormen. The whole aspect of the road has changed with many new high rises. I find it difficult to envisage the road as it was. Merry Court, a then-enviable new building that came up to block our view from our balcony, now looks very used and in need of updating.

Living on Robinson Road had its joys. The Botanical Gardens were nearby, which provided some greenery and nature for us. We missed the open spaces and public parks of London. The Caine Road offices were within walking distance, as was the school for my daughter. Her friends used to think I was her elder brother. With my long hair and sandals I looked very young. Robinson Road was a treacherous thoroughfare. The pavement was narrow, traffic was heavy, and dogs were walked daily to do their business as their owners ignored the mess left behind for unwary passers-by. This last item has not changed much in the thirty years since.

At the Robinson Road home Gwen learned to cope with her first maid. Actually the maid chose her, even though normally the employer does the vetting. Unsure of how to hire or fire domestic help, Gwen, with her then-inadequate language abilities, hesitantly stood outside an employment agency composing mentally what to say. As she nervously stood there on the sidewalk, a jolly, plump lady was studying her.

"Are you looking for a maid? I think you look like a nice lady, sincere and honest. Not likely to cheat or beat me. I am hard working and experienced. We would both save ourselves a commission to that thieving lot. Why don't you try my services? We can always part ways if it does not work out!"

That was how we acquired the help of Ah Geen. She looked after us and fussed over the family like a mother hen. Widowed when her youngest son was just a few months old, she had struggled to rear seven children on her own. Having a home nearby in Hollywood Road, she opted not to live with us. Unaccustomed to having a nonfamily member around for twenty-four hours every day, we were quite content for her to leave after dinner. Her young son was only six and needed his mother.

From Ah Geen we learned the customs, mores, and local values of life in Hong Kong. I was supposed to bribe the policeman who ticketed me for not stopping at a stop sign taken down during a typhoon. Gwen should know how to check if eggs were fresh. One holds them up to the light of a bulb provided by the stall keeper for that purpose. Gwen never actually quite knew what she should look for, but went through the motions. Thankfully, the fledgling supermarkets began to appear in the late 1970s and by the year 2000 they were comparable to any of the largest in the West. Many locals however still prefer to do their food shopping in the neighborhood markets. In contrast to the hygienically packaged fresh meats, fruits, and vegetables, they prefer the advantages of being able to poke, prod, and pick their purchases.

The teenage daughter of Ah Geen, Bicky, her sixth child, was enlisted for a small fee, to coach our children in Cantonese. Slowly we were all beginning to adapt. Gwen even felt sufficiently confident to restart her temporarily abandoned career. She joined the message switching division of Cable and Wireless. Being bilingual was an advantage then, as it is now. However, in 1970 it was rare for a Chinese, apart from the very rich, to be as adept in the colloquial usage of English—so she saw two sides of the working world.

"Why," she asked her English boss, "do you openly display the payroll sheets for all to peruse and sign every week?" "They are like children, you know. They get a kick out of seeing it all!" he laughed.

The giggles of the employees as they signed were of embarrassment.

"This stupid *gweilo* has no technical know-how of this business but we humor him and pretend we are idiots!"

Neither side understood or privately respected the other. That was colonial Hong Kong.

Sadly, Ah Geen was not able to remain in our service when we moved to live on campus out in Shatin. It was too far from her home. By the time the family packed up to live in campus quarters, the children had suffered a number of changes in schools as well as homes. Simon was not doing well at the Chinese school. We needed to change to the Kowloon schools sooner or later, if we were to live on campus in Shatin. It would be easier if both children attended the same school and they could watch out for each other. We again approached the ESF schools. Again we were answered in the same tone as with our earlier attempt to enroll the children. This time we fought back. "What is wrong with their English? They are native speakers. They attended schools in the UK. English is their first language"

Hong Kong has an interesting and aggravating attitude as to who is considered a native speaker of English. One with a Caucasian face, however appalling their English might be, is preferred over an Asian with a perfect grasp of colloquial English. It baffles me to this day as to why the Education Department continues to maintain that only a person with a Caucasian face is able to speak native English. Gwen does not even qualify. She verified this fact when she applied for a vacancy as an English tutor at a school, thinking she was doing the community a favor. At the interview they responded that she was not suitable for the job. Gwen had not even opened her mouth yet, but she had the wrong facial features.

The school insisted on an English test for my children. The ESF School at Beacon Hill could not fault the test results and with reluctance and much grumbling, they accepted them as students. Immersed once more into a schooling system with which they were more familiar, our children thrived. "We sit at tables." Amanda cheerfully informed me. "I am at the top table."

"How do you know that? Is it called Table No. 1?"

"Oh no, we are table C, but we all know which is the top and which is the bottom table."

I was a little cross when I discovered she was also teaching slow learners to read. The school, which had questioned her abilities in English, was using her as a teacher's aide. In the UK she had learned to read early and at eight was a proficient and avid reader. Simon also was happier. As they were now in an English-speaking environment at school the progress in Cantonese was unfortunately slower.

The contractors fell behind on their building schedule at the university and we were forced into another short stay in a hotel. The Shatin Heights Hotel, now long demolished, was then followed by a six-week stay in an apartment sublet. The residents of this address were the family of a Cathay Pacific Airlines pilot who were in the UK on a long leave.

I do not recommend our way of life during those two stressful years of transition. Though one good thing that did come out of it was a very close-knit family relationship that has lasted to the present day. The children, in their frequent distressing loss of school friends, adhered to their parents for support and activities.

We taught them the rudiments of tennis and board games, read them stories, went on hikes in the hills, made weekly visits to the public libraries and, as non-swimmers, bravely plunged into the university pool to learn to swim together. We tutored them in their homework, supervised their piano lessons, and accompanied them to cooking classes at the YMCA. They had their moments of rebellion, such as crashing on the piano keys as loudly as possible at 7:00 a.m. to express their dislike of the instrument. Simon had a spectacular spill on his bicycle riding down the treacherous steep hilly roads of the campus. Away from the strict eyes of his parents, he ate forbidden foods from the open-air food stalls. He learned all the curse words from his school friends. Of course, his parents were only partially aware of these adventures, many of them being related to us years later. On top of this Amanda was nearly swept out to sea at summer camp one year. She and another child shared a canoe paddling in the choppy waters off the Wu Kai Sha shoreline. The waves and winds were pushing them outwards to the rougher seas as no one noticed the danger.

We finally were able to move into the Residence 1 staff quarters at the end of the summer in 1972. I can see the same six-storied building in the haze from the heat in the distance as I wait at the train station to visit my doctor.

The minibus swerves to a stop and I board it with several other waiting passengers. As I walk across the road I feel as if I am walking back in time. The banyan tree is still standing majestically on the side of the grassy lawn in front of the apartment building's main entrance. It has grown quite a bit over the thirty-plus years. On the far side of the grass are two stone benches. The same benches that a passing family from the city, on a Sunday outing to the campus long ago, had thought was a good place to settle on for a barbeque dinner. The security was called out to remove them as they lit up the charcoal and very noisily disturbed the peace of campus life. There, too, is the concrete sidewall against which the children and I had hit numerous tennis balls, much to the annoyance of the other residents.

As I walk through the wooden doors that are no longer immaculately varnished and shiny, the entrance lobby area looks grimy and well used. I look at the elevator door, scratched and dull, where once an anonymous letter had been posted: "Please stop piano practice in the early morning. We find it disturbs our sleep."

The notices now on the elevator door inform me that the medical offices are on the first floor. I turned to walk up the stairs. Our apartment had been on the second floor. In all the years that have passed since my home had been in this building, I had not once set foot into it, though I must have passed by this building many times during my years as vice chancellor.

One year, Gwen had organized a summer project for the children. Our two children and two of their young friends spent some weeks making papier mâché puppet heads to fit over their small fists. It was the sort of messy fun that appeals to young minds. This task consisted of tearing up old newspapers into pieces and soaking them in buckets of cold water. Then using plenty of flour glue, the puppet heads were molded by small hands into the characters of the story *The Three Little Pigs*, dried, and then painted with poster paints. The children composed the dialogue for the well-known story and rehearsed it all by themselves with some directions needed from the adults. It kept them fully occupied over the long summer school vacation. When they were finally happy with their play Gwen put up the flyers, also designed by this amateur theater group, to invite all the children on campus to a performance in our home.

The wooden folding doors, which divided the study from the living room, were pushed back and a long three-seat sofa rolled in front of the opening, facing inwards towards the study. The little performers crouched on the seat and waved their puppet heads above the high back of this sofa as a makeshift puppet stage. The audience of about ten little children and their chaperones were quite enthralled. Pauline wept so realistically as her house was huffed and puffed on by the Big Bad Wolf that her mother thought the tears were genuine tears of stage fright. That time seems so long in the past, such a distant memory. Now I am old. I realize how short time is when my children were a close part of my daily life.

The long corridor leading to the bedrooms faces the front door. If the door to the toilet at the end of the passage way is inadvertently left open, as it often was, visitors' first view of our dwelling was a direct line of sight from the front entrance to the white lavatory. As I walked through that passage to the consulting room of my doctor, I saw the same white lavatory. The door of the small room is left ajar though there is a notice pinned to it now: "PLEASE KEEP THIS DOOR CLOSED."

I cannot recall how the bedrooms were anymore or indeed which child occupied which room. After the move into Residence 1 we greatly missed Ah Geen. As Gwen was working in town, traveling back and forth on the train daily, she relied on me to rush home from my nearby office in emergency situations. In turn I relied on the maid to be sensible and capable.

From the contacts of the Cathay Pacific Airlines family, we had hired a new maid, Becky. She turned out to be unreliable, arriving late for duties. She was a dreamy girl in her early twenties. One day she simply did not arrive at all. It was during a school holiday and Amanda telephoned her mother: "Mummy, Becky isn't here and it is lunchtime. What shall we do?"

We finally managed to locate her family only to be informed that Becky had quit her job: "She got a job in a factory and does not want to work for you any more!"

No thought of giving decent notice had crossed her mind. The new job must have paid so much more that she was happy to forfeit her pay. It was about that time in history that local maids became unattainable. Women were in high demand to fill job vacancies in the many factories sprouting up across Hong Kong making various electronic gadgets. That was before these jobs moved over the border into China and Hong Kong lost its manufacturing labor pool. This marked the beginning of the labor influx from the Philippines.

We went into panic mode as we needed a replacement fast. A handyman at the office offered the services of his elderly aunt. She moved in and turned out to be a grumbling person full of complaints. She would say this was wrong, that was not good enough; only brand new bedding was good enough for her use. The children were too noisy; the dirty laundry was too much to wash. Never employ someone's elderly aunt.

The final straw came in the middle of a terrific typhoon. During the night, strong winds shattered one of the windows with a big bang as the pressure of air was released. The howl of the wind entering our home whistled like a frantic genie let loose. The children screamed in fright. The rain flooded in. I grabbed towels, anything to soak up the water. I found material to temporarily block the hole, banging in nails loudly. The elderly aunt "slept" through it all and emerging from her room in the morning, claimed not to have heard a thing. "All these wet dirty towels for me to wash. All this mess I have to clean up!"

For Gwen, that was it. "You are fired. This instant. Just pack your things and go. We have been up all night battling the elements with no help at all from you and you complain!"

We were more careful when we vetted the next maid who had come recommended by her previous employer at the University. Ah Fen turned out to be as good as Ah Geen. She stayed with us until 1974, when we left Hong Kong to go to the United States.

I walked into the temporary consulting room of my long-time medical practitioner. She had joined the medical department some time before I became vice chancellor. Gwen likes to consult her but I used to see the head honcho. When he retired my wife persuaded me to visit her doctor, and I have been happily doing so ever since.

"I hope you have not eaten breakfast," my favorite doctor says, "because I need blood samples. Let's see, it's been over a year!"

Alas, I have eaten, so I make another appointment for the next day and after some pleasantries walk down the stairs and out of the building. I pause at the door and look down the hill at the view in the distance.

Below I can see the old sports stadium in front of the railway station. Beyond, on the skyline across the bay, is one tall high rise beside another—the Ma On Shan town center. The strip of water that is the bay appears like a sliver of silver in front of the town. On this side of the bay sits the Tolo superhighway behind the train tracks. It is traversed by heavy container trucks in both directions.

In the 1970s, Gwen made her daily commute to the city, she had stood on the train platform watching the sea gulls swoop down to catch fish in the waters of the bay lapping just a few feet away. The canoes for hire bobbed up and down gently with each wave. The only way to cross to the other side of the bay, apart from the long drive around it, was to hire a *wallah-wallah* boat. I never knew why this name came about. Perhaps one hailed it by shouting out "*Wa, wa la!*" It was an idyllic scene compared to the present setting.

Before, the only road in and out of the city ran mostly parallel to the railway line. At one point it went over the tracks at a crossing. The children loved being held up at this crossing as they waved frantically at the train when it passed by our car with a swish. The drive into the city was only a little faster than by train. There was only one two-way tunnel through the mountain ranges and we made the harbor crossing to the island on car ferries.

The saddle of the mountain over Ma On Shan still dominates, but the small villages and the shorelines of Wu Kai Sha are gone forever. My eyes wander up the ridge and around to the top of the steep mountain.

Did I really climb to the top of the saddle? The young students of the university climbed it as a yearly challenge, racing to the top to break previous records. One year my family and I joined them, crossing the bay in a *wallah wallah* boat. This was a largish flat-bottomed boat for about four to six passengers and its owner, usually a wizened old lady, sculled the boat across for a small fee. It was a scary crossing on the low skiff, especially as the students overfilled the boat's capacity in their rush to cross quickly.

The ridge trail narrowed higher up the mountain and I do not re- member getting to the top at all. Perhaps my son did. I remember passing by the old abandoned silver mine and its rickety wooden mineshaft. The students now no longer race up the mountain. In time this traditional climb has been abandoned and replaced with other challenges. One of the new traditions is the annual crew race from each university in Hong Kong. This takes place on the river that runs into the bay. The initial ad- vantage CUHK had, of being so near to the practice waters, has gradually dissipated over time as distances became shorter. Where once a trek to the New Territories was looked on as if traveling to a distant country, the improved road infrastructure and the speed of modern modes of travel have made the area seem as if it is just next door.

In 1997, when Hong Kong reverted to China, one of the proclamations made was that certain conditions would remain unchanged for the next fifty years. I find this a little unrealistic. The changes I see in Hong Kong after thirty years are tremendous. Even the mosquitoes that bite me are of a different breed. Changes happen and no one can stop its wheels from turning.

I walk down the hill past the Chung Chi Staff Club, past the Chapel, where I am stopped by an old retired academic. He, like me, is a visitor on campus—relics of a previous era.

If anyone had prophesied to me that when I left the university in 1974 I would return one day as its president, I would have said that it was impossible. The fact of going to America to continue my work in the advancing fiber optics industry was already a dream come true. I felt lucky beyond expectations.

During my tenure as vice chancellor at CUHK, my father arrived in Hong Kong in time to celebrate Christmas with the whole family. My children were delighted to reunite with us for a short time, flying over from the U.S. Both of them were in the first stages of a career in the technical world, and were not yet interested in finding work in the East.

As the suite of rooms was quite large in the residence, I had a desk and bookshelves put into my father's bedroom. There was a large built-in wardrobe with drawers, with a pantry for making tea and a spare room that he could use to meet visitors. His quarters were more than adequate. After complaining that his quarters were too small, he eventually accepted it and settled in. Gwen made arrangements for his care. We had a cook, a maid, and a driver. With Gwen's stern approach, the maid oversaw his bath, which he was forced to take, as a compromise, twice a week rather than daily. He met his match in Gwen, who yelled at him if he did not obey.

"Daughters-in-law are supposed to love and respect me," he complained to me during my five-minute visits each morning. He liked to write out shopping lists over and over again, never discarding the old ones. Gwen sorted that out by casually reducing the piles of old papers every now and then without him noticing anything was missing. Father was nearly ninety years old and extremely healthy. His appetite was very good. He enjoyed his food enormously and would overload on the richer foods with a nonstop snacking of nuts and chocolates between meals.

"Drink," I encouraged him, "Let's see who finishes the glass of water first! Me or you."

He had gallstones for which he had to be hospitalized. There was a repeat of his past behavior in hospital. He was as difficult as when he had suffered his cracked bone twenty years ago. He groaned as loudly and we had the same complaints from the hospital as before. "Can you please curb your father; he must not ring the bell incessantly for the nurse. He really is disturbing the whole ward!"

I thought at first that he would enjoy going to the library at the university. However, he caused such a disturbance there, demanding attention loudly, that I had to forbid it. Then there was the hilarious time he went missing. As strong as a horse, he was able to slip out of the garden gate, unnoticed by the gate guard, and walk the mile-long drive downhill to the main road. Gwen was out of the house and the maid, not finding him anywhere after a search of the grounds, raised the alarm.

Security was called and questioned. Someone said he saw an old gentleman, smartly dressed, waving down a car right by the campus exit gateway. He thought the car belonged to a staff member whom he could name.

When we eventually tracked down the staff member he told us that an old man had asked for a lift into Shatin. The driver had dropped him by a bank that the old man had pointed to. While all this investigation was going on, he was at the bank arranging for a transfer of money to a "needy nephew" in China. I had forbidden him to do this, as he needed his money more than the nephew. Besides, he practically had no money left to give away. He wanted to feel like a king again, providing handouts to the world.

I was quite worried as to what could have happened to him. A taxi rolled up to the house and he triumphantly emerged. "You are lucky I picked him up and let him get into my taxi," complained the taxi driver, angling for a big tip. "You should keep a better eye on him."

Father was very unhappy with his loss of independence. I was forced to have a slatted half door installed in the archway to his rooms. He could see out but the door could be locked. It thwarted him from wandering at liberty around the house and rifling through my papers. Sadly, his world was constantly shrinking. He lived with us until almost the end of my tenure at the university. In preparation for leaving my post, I took the advice of an old friend and arranged for him to live at a newly opened home for the elderly near Hong Kong University. After my retirement, I was not sure whether I would still be living in Hong Kong. At my farewell dinner, I had been evasive as to where I would go next. I really did not know. Gwen gave a clever answer, and everyone listened with bated breath and great curiosity to see if she would reveal anything.

"I know very well where I will be when my husband retires next week," she announced. "I will be right at his side!"

Most people at such a senior rank would have had private conversations with their many contacts in the business world, and probably would have had numerous job offers to consider. It never occurred to me to do this. I planned to give some lectures as a kind of open-ended vacation around Asia and then perhaps move closer to my children. Maybe I would putter around and write. I was very unsure as to how I would occupy myself to fill the long hours of leisure time. Of one thing I was sure, I would be very busy doing something.

MR. AND MRS. KAO AT UNITED COLLEGE ON THE CUHK CAMPUS (2008)

Wedding Anniversaries

My wife and I celebrated our forty-second wedding anniversary in New York. Friends had suggested the Peninsula Hotel near our apartment as a good place for afternoon tea. So that's where we spent a couple of hours on September 19, 2001.

Today we constantly hear about short-term relationships and divorce and very little about long-lasting marriages. My parents were both surprised and disturbed to learn that I had known my then-girlfriend for only two years before proposing to her. Over that time we had become great friends, and sex was just a distant fleeting thought, though the chemistry was certainly there. The problem with today's relationships is that sex is foremost and friendship is the afterthought. Friendship between two people is the glue that holds them together for decades. In this final stage of my life, when desire is muted, to have a best friend with whom I can share common interests and confide my innermost thoughts, feelings, and desires without any inhibitions is as essential as breathing. How lonely would life be otherwise?

With the terrible terrorist attack on the World Trade Center in 2001, it was an especially significant year to spend quiet time together and to be thankful for our lives. On that terrifying day in September, I was having breakfast as I watched the early morning news. The events evolved live as the cameras switched immediately to record the awful happenings in Manhattan.

The TV studio was in a building where the World Trade Center towers formed a backdrop for the newscaster. The sharp eyes of the cameraman saw the impact and explosion on Tower I. Horrified, the commentator initially yelled—"a terrific explosion, oh my god . . ."

Then they continued to run the footage again and again—in slow motion. Was it a plane that had struck the tower? Was it engine failure? Had it lost control? It was a small commuter plane . . . no! It wasn't! The commentator was as confused as the viewers. How could such a disaster be happening? Why would a plane be flying so low over the city?

Heavy black smoke billowed up from the point of impact. Fifteen minutes later, as the commentator was still unable to confirm the exact cause of the first explosion, a bright orange flash ballooned out from WTC II at a lower level.

As I watched the images, I noticed a plane coming across from the river behind the towers. When the camera reran the footage of the second explosion, the plane's wing distinctly appeared behind the tower just before impact. Cameras from several other angles focused on the scene. They all confirmed what was now obvious—two planes had hit the towers like missiles. This was not an accident.

We never thought of going outside to look down 5th Avenue to see and hear the destruction with our own eyes; we were glued to the TV for the rest of the day. On TV it felt like a movie—an impossible scenario that could not be real.

The second tower collapsed first, although it was hit fifteen minutes later. They surmised later that this was because the lower point of impact and the intense heat from the burning jet fuel had melted the central core. With the heavier weight of the floors above the damage, WTC II fell an hour after the assault. WTC I sank down more slowly about a half hour later. With unbelieving eyes, I watched as the tall antenna on the roof, fighting to remain upright and vertical, dropped lower and lower over many seconds before the dust storm eventually obliterated it.

I had heard the sirens from the fire trucks, the police cars, and the ambulances racing by. The crews scrambled into the buildings to rescue those trapped inside. What those first firemen and police on the scene didn't know is that they would have no chance of escape before everything would tumble down into one gigantic rubble heap.

Our conversation was philosophical. What would the U.S. government's response be to the terrorist threats of a faceless enemy? We both had experienced wartime strife—in Europe and China, respectively. Would this lead to a third world war?

More than a week after the attack, neither of us felt any desire to go and gawk at Ground Zero. Gwen had lived through the London Blitz of WWII. She had seen and tasted the acrid smoke from the bombing raids by the enemy; she had played in the ruined shells of what were once family homes. In childish innocence, she had no knowledge of the buried bodies and the human tragedies. Now the thought of the lost lives and the pulverized human remains buried under all that weight of the destroyed towers horrified her. We both knew the world would no longer be the same.

Though still disturbed by the unfolding events, we celebrated our anniversary tea with a marvelous spread of tiny savory sandwiches, hot scones with cream and strawberry jam, and a variety of interesting pastries. "My best ever anniversary gift to you," my wife says, "was our daughter."

Amanda, our second child, was born at 3 a.m. on September 19th, on our fourth wedding anniversary. She was obstinate from the start. She kicked fiercely, announcing her imminent arrival by breaking Gwen's water, but then changed her mind as she made her mother stay in the hospital for another uncomfortable day. Nothing seemed to be happening after the initial action, so Gwen, miserable in her hospital bed, decided that she should go home. The attending midwife disagreed, even though she thought it would be many hours before the birth. She sent me home to bed instead. That's when Amanda decided she would arrive in a hurry, taking everyone, including the stunned midwife, by surprise.

I had left the ward a little after 11 o'clock that night, leaving behind a tearful wife who wanted to go home with me. "Here, take a couple of sleeping pills and go to sleep," the nurse said to Gwen, "You are just being fussy!"

Half an hour later, she summoned the nurse again and there was no mistaking that the birth was indeed imminent. Amanda arrived just over three hours later—an unexpected and happy surprise for us all! At 8 in the morning when I dropped by the hospital on the way to the office,

I had no idea that I was already the new father of a baby girl. It was a very exciting day to celebrate two events. There is nothing like that sense of elation and joy from holding your newborn child in your arms and beaming at the wonder and miracle of life.

Along with our two-year-old son, the following months were full of the struggles of raising two small children. I would come home from the labs to find a wife under stress and overwhelmed by the day's housework, cooking, and washing. Home help was not affordable, and not readily available at that stage in my career. As new parents we had to rely on ourselves—my family was far away on the other side of the world. Gwen's relatives had problems of their own. So I would change, bathe, and play with my son in the bathtub while Gwen cooked dinner. I was the first of the modern fathers! I even wheeled the baby out and about in her pram— a no-no for macho men of the era.

Family life was fairly uncomplicated, but very busy. Mondays to Fridays I went to work at the research lab and exercised my thoughts on the intricacies of rectangular wave-guides and, later, on the theories of optical wave-guides. Outside of my working hours, I helped out around the house. I hugged and played with my children, romped on the floors in mock battles with them, took out the trash every night, hung out the wet clothes to dry in the sunshine in the garden. Clothes washers were still a novelty and our first expensive purchase was a Parnell. Rather than spend more on an electric clothes dryer, we decided to continue to harness nature and use sun power. That did not come in great quantities in the UK and our home was often draped with damp, drying, cloth nappies.

I suppose my colleagues might have viewed me as a wimp. Men then were not encouraged to do the things I did in the home as second nature. The men of today have mostly caught up with the times, which makes me less of an anomaly. How much easier it is for today's parents with a vast array of disposable baby care items and affordable appliances at their beck and call.

Shops in the UK used to close for business at 5 p.m. so any shopping was out of the question during the working week. Hence, the majority of my

Saturdays were spent shopping for the necessities of life. Sunday, supposedly a day for relaxation, was either used for catching up on house repairs and housework that had not been accomplished during the weekday evenings, or driving westwards an hour or more across London for the obligatory visit to my mother-in-law. Getting two small children and all the paraphernalia to go with them was a time-consuming task and invariably we arrived only just in time for lunch. This meant we could not leave for home until late in the afternoon. My mother-in-law was never satisfied unless it was a whole day's visit and usually verbally attacked us for not staying until after dinner.

"Why do you want to leave so early? You may as well stay for dinner now. You don't like my cooking! Not good enough for you? Why do you bother to come if you want to go so soon!"

So we would unhappily agree to stay longer and sit thinking about all the small urgent tasks we could have been doing in our own home. My mother-in-law had no life of her own. She lived her life through her son, her favorite, and when he was not available, through her daughters.

Such Sundays were a revolving drama. Gwen only came to understand her traumatized childhood after decades of reflection—her mother, who always seemed more than a little schizophrenic, was most likely mentally ill. The repercussions of that childhood affected my wife strongly all through the early years of our marriage. It required much understanding and patience for me to help her get through the rough patches—or maybe it was my lack of understanding. Much of my mother-in-law's ravings luckily went over my head as she spoke a village dialect beyond my comprehension. I simply smiled and nodded.

Decades later when I met up with friends from my university days, they wondered what had happened to me: "We heard you got married. But then you disappeared and never contacted us again!"

By the time my daughter was six months old, Gwen was determined to find a role outside of the house. The research lab offered her a part-time job, but we needed to find reliable help for the mornings she was away.

Our fifth anniversary was a working day at the office. I had begun to experiment on the possibility of optical fiber systems, while Gwen used FORTRAN programming to find solutions to intricate mathematical equations. Some of these equations were the very ones created by me and my colleague, George, which later become a part of our 1966 paper, published by the Institution of Electrical Engineers in London. Gwen recently remembered this work and pointed out how I never gave her credit in the paper.

The weeks, months, and years went by too quickly, as did the anniversaries and birthdays. The latter being of greater importance, as we took a back seat on the 19th of September for many years. The day was for celebrating Amanda's birthday, and for Simon's birthday, too, as that came only three days later; he had been a late second wedding anniversary present. Life followed a fairly regular, busy routine during the sixties. At the time we thought it would never change and that we would grow old together in a home in England.

In 1970, an unexpected offer came. Friends from long-ago university days in London contacted me. A university in Hong Kong was setting up a department of electronics from scratch, and would I be interested in getting it started? I gave it much thought and discussed it at length with Gwen. Exploring the possibility of taking a sabbatical leave of two years from the company provided a less risky move. I could rent out the house and therefore not cut off our return route to the West.

In England my family and I were an ethnic minority, and I was beginning to feel the pressure of the glass ceiling as I progressed upwards in my career and research. Returning from meetings in London late at night, I was getting rather tired of the same questions from the taxi drivers, who all assumed that I was a waiter in a Chinese restaurant or the dishwasher.

Growing up in England, my son was also beginning to question his identity: "They make fun of me at school and call me flat nose. Why can't I be English? Why do I have to be different? I don't want to be Chinese!"

For all of these reasons and many more I had not yet thought about, I chose to accept the offer in Hong Kong. And so our eleventh wedding anniversary involved a big move across oceans. We arrived in September and spent our first month in a dimly lit room in the Cathay Hotel in Causeway Bay, a hotel now long demolished. The birthdays and anniversary were left uncelebrated in the strangeness of our surroundings. I was no longer conversant with local manners and customs. I needed to relearn how to order at restaurants, request my bill, and back stab on the job in the Eastern tradition. In short, I was more used to Western ways and habits—even speaking English came more naturally.

We extended our original two-year term for an additional two years, which meant that I was able to oversee the first batch of my electronics students through to their graduation. The initial intake of students and lecturers that I recruited was small that it felt like a family. In fact, at the time of its development, the whole university was like one big family. The student population numbered about four to five thousand. The size of the academic staff only ran in the low hundreds and we were all on nodding terms with each other.

I had become accustomed to the more casual culture of a research lab. I was reluctant to wear a suit and tie in the hot humid climate of Hong Kong, especially since many of the facilities lacked air conditioning. I became somewhat notorious for being casually dressed, without socks and in sandals. Many people even mistook me for a student.

After four years away from my time in industry, instead of returning to the lab in England, the ITT Corporation invited me to transfer to their U.S. plant in Roanoke, Virginia. Optical fiber development was progressing from the drawing board to pre-production of a working transmission system.

So we spent the next eight wedding anniversaries in Virginia. During those years, my peers began to recognize me for my pioneering work. With increasing financial stability, we were finally able to celebrate these special occasions by dining out or going to Washington, D.C. to visit the marvelous museums.

Life in America was an entirely new experience. Rural towns and small cities across the U.S. are very insular. At the time everyone seemed only to read the local papers and had little interest in what went on in the rest of the country, let alone elsewhere in the world.

When I told my next-door neighbor that we were driving three hours to visit the museums in the Capital, he thought I was insane. "Why do you want to waste your time with all that driving? We have museums in Roanoke and everything else anyone could want in life! I've never been there and I never will go. Lived all my life here and it's good enough for me!"

The U.S. economy in the 1970s was strong and the Roanoke plant, marketing their production line of night vision goggles, showed ample profits. Soon after arriving in 1975, a few of us top managers, together with our respective wives, were invited to a three-day meeting at the Greenbrier Resort—a top hotel in West Virginia. Everything was new— the culture, the way of life, even the food that came in quantities to feed an army.

Every year, the Greenbrier trip was held around the end of September, so it became our private wedding anniversary celebration. We worried about leaving the children on their own and asked neighbors to drop by on the first occasion we left them at home alone. Gwen phoned them every night. Once we knew they were fine, we quickly learned to enjoy the American way of high society life with nightly dancing and drinking, all dressed up in black tie and long gowns. The formal dinners and buffet lunches were fantastic. The Greenbrier also ran a training school for gourmet cooks, so top professionals in the trade prepared the meals. My colleague, Jim, and I competed as to the number of times we could visit the dessert table.

While the wives spent the days lounging by the pool and being beautified, the men were occupied with work. On the one free afternoon, most of the men went off to play golf. Jim and I had never played, so another colleague, Bob, offered to coach us. Bob declared himself an expert at the game. I tried hard to hit the elusive, tiny, white ball just as Bob taught us, but found it more fun driving the golf cart. Gwen watched on the sidelines and laughed so hard that she collapsed into tears. My first golf ball hit the cart. Jim managed to hit his ball up into the branches of the tree next to him. Bob walked off in disgust. He did not want to be seen in the company of such duds.

When I signed my bill at the end of that first trip, Gwen gasped at the amount—even though the company was picking up the tab. "Check the figures again." She said, "They must have made a mistake!"

We were less gauche the following year. Knowing the ropes allowed us to relax and to enjoy the occasion. We felt young and carefree again. That year, our seventeenth year of marriage, the children insisted that they were old enough not to need the neighbors checking up on them. They proved themselves by sensibly phoning me at the hotel when a telegram arrived from the UK informing me of my mother's death. I kept the news from my colleagues, as it would not have served any purpose to leave immediately. The funeral was to be a week or so later and my father had already made all of the arrangements. That muted our memories of that particular wedding anniversary. Though my wife and I attended the third retreat, the novelty had worn off. Besides, some of the wives had been divorced or otherwise moved on. A sense of gloom seemed to hang over the event. One of the now-divorced couples had been wonderful dancers, a delight to watch on the dance floor. Now Maria had left her husband, unable to stomach his constant womanizing. So he claimed my wife, who is also a good dancer, as his partner one evening and danced her off her feet. She became very embarrassed, but could not get away until she feigned tiredness and retreated to her room. There was no fourth Greenbrier workshop and we returned to our normal dinner outings to celebrate our anniversary.

I did not know then how grateful I would be for these introductions to formal entertaining, black-tie affairs, five-star hotel living, and all the necessary etiquette that went with it. The experiences served me well in the decades to come. My wife also grew in confidence and it gave her the poise needed to accompany me as my career took off. I am grateful to the ITT Corporation for many things—even though they retained all of the rights to my patents.

I had expected to settle in Roanoke for many years more. By then both of the children had graduated from high school and were away at college. Alas, the custom home we built was destined never to become a home for the family. A few months after moving in without the children, I came home from the office with an announcement: "I think you should sit down for this. I have good news and bad news. Which do you want to hear first?"

She opted for the good news first, which made the bad news obvious.

I had been appointed an executive scientist—the first such post in ITT history—with the option to choose where I wished to relocate. I chose to leave Roanoke to go north to the research labs in Connecticut. We put the new house on the market and then I left my wife to fend for herself. I was off to seek more fame and fortune on another stage. That 19th of September we were apart and we missed each other terribly, especially as we were now empty nesters.

During such transitions, the working spouse always has an easier time settling into new routines and making new friends. The excitement of different responsibilities, interaction with a new supporting team of colleagues, and the intense mental activities make the adrenaline flow. The days pass quickly. It is more difficult for the nonworking spouse, especially if there are no longer small children at home or a way to connect with the neighboring families.

Initially Gwen had not been able to work legally in America. In Roanoke, a very conservative community, women were expected to be full-time homemakers. Any other role was seen as an inability of the husband to provide for his family. As one of the early pioneers of the 1960s in computer programming, Gwen could have been offered a position with my company, too. But they had an antiquated rule of not hiring employees' spouses.

However, being very resourceful, she found mental outlets through a variety of occupations, including as a temporary substitute teacher for a while when a friend at a local school took leave. Teaching trigonometry to a mixed class of 12th graders was no easy task, especially as Gwen had not picked up a math book since her college days twenty years ago. She would read the textbook one page ahead of the class to stay just in front of them.

After substitute teaching, Gwen tried her hand at volunteer work. This entailed stuffing envelopes and composing the business letters for a boss who was not very adept at writing. Finding the task extremely boring, she went back to school and drove three days a week to Virginia Tech in Blacksburg, thirty miles south. She found the courses fun until she signed up for the auditing class, which entailed the memorization of numerous rules and regulations. The C grade depressed her and the final straw came when she had to follow around a young man half her age during an internship at a local accounting firm. He expected her to make the coffee and he read aloud to her from textbooks as if she were incapable of reading herself. Finally, she and a partner, the wife of a colleague, bought a dress shop and ran the business together for two years. They hired the partner's three daughters and our two children as part-time staff and learned how to cope with customers, take inventory, arrange the goods in the store windows, keep the books, and battle shoplifting.

When our daughter left home to go to Duke, Gwen decided to enjoy some hard-earned free time and drive down to visit or collect her from school. So when I left home too, she moped alone in her beautiful home, packing up the household belongings in silence. My phone calls home full of my new adventures caused her more pain than I realized.

"I went out to dinner again tonight with Elaine, a colleague at the lab. It was a lot of fun. It was to the same restaurant and the waiter there must think we're husband and wife! She is married to a guy who is out of town on business. She says he doesn't mind her having dinner with me." I continued, "Wait until you come up—you'll like her. I'll introduce you!"

Women have a sixth sense that men lack. I was full of naïve enthusiasm about my new friend, while my wife imagined disaster and cried herself to sleep every night. Much to my surprise, when Elaine did meet Gwen they were very cool to each other.

It was hard to give up a custom-built home without having the chance to enjoy it. So we ventured again into the thorny thicket of finding a contractor, a piece of land, and a new house design. The building of this new home in Connecticut took the usual amount of time, so Gwen was stuck in Salem for longer than planned. There was a lot of driving between there and my rental apartment in Shelton. I think we had lobster dinners by the shore for our 23rd wedding anniversary.

In 1984 the local ITT office in Puerto Rico requested that a top corporate ITT representative come down to give talks about the new optical fiber business opportunities. "Why not!" said my wife, "We can celebrate our 25th in grand style."

The week before we were due to fly to San Juan, I slipped on the grassy slope of the backyard. I had been clearing large stones from the bottom of the yard and moving them up the slope to the front, where I was planning to make a rock garden. The wheelbarrow, full of large stones, toppled over, and took me with it. One of these stones bounced upwards and hit me in the left eye. The next day, I sported a black eye as large as a small plate. Gwen said I looked like a gangster.

The official party meeting us at the airport in San Juan might have known that they would not be meeting the usual type of American businessman. They were very polite and gracious. They presented a large bouquet of flowers to Gwen and welcomed us in Spanish. Over the next few days, they drove us around town to meet various leaders of education and government, fêted and dined us in between my lectures. Bodyguards accompanied us everywhere we went. In our free time we did some sightseeing, though it was difficult to understand since their English was rudimentary. We had a great time.

The final day was September 19th, and we reserved a table in a nearby restaurant that the hotel concierge had highly recommended. After dinner, we wanted to take a romantic stroll in the twilight by ourselves, but the bodyguards vetoed the notion and insisted that it wasn't safe.

The guards drove us to the restaurant, dropped us off at the front entrance, and said that they would be back in an hour or so. The dinner was delicious and we savored it, taking our time. Twenty-five years of happy marriage is indeed something to celebrate, especially since the next day we were flying home, back to our routines. We were the last people to leave the restaurant, and when we finally exited we found ourselves in an absolutely deserted, pitch-black street without a car in sight. We turned back to re-enter the restaurant and found the door locked. Where were the bodyguards? We were certainly easy targets for any would-be kidnapper, which is something we had in the back of our minds, as there had been a spate of American businessmen kidnapped in Latin America in 1984. Banging loudly and shouting, the waiters finally unlocked the doors. We were supposed to have exited out the back into the car park. The bodyguards had been waiting there all along.

Weeks after our trip, we received a large envelope from ITT headquarters in New York that contained glossy photos of our trip. The "Asian gangster" with the black eye in San Juan had been permanently recorded for history.

In 1985, we quietly celebrated our twenty-sixth wedding anniversary by ourselves in Germany. We probably went to dinner at the restaurant in Gerlingen, a nearby small town, because they served wonderful food. This was also where we learned that the first helping of the main course comes again as a second helping in the same large amount. Gwen made great efforts to learn the proper, high German, as she had been quietly told that it would be a mistake to speak broken German like the despised foreign imported workers. She succeeded reasonably well, enough to make a tolerable farewell speech for me at our goodbye party at the end of the year. I did not need to learn the language as everyone who was working with me in the labs spoke English, whereas Gwen needed to go shopping and socialize.

In 1988, I was elected as a foreign member to IVA, the Swedish equivalent of the National Academy of Engineering in America and the Royal Society of Engineering in the UK. This was a great honor for me. They held a big annual event for its members and new foreign members were invited to attend when they were elected. I might not have seriously considered attending the event, but Gwen encouraged me to go: "It's our twenty-ninth wedding anniversary this year and we have not planned anything. Wouldn't a trip to Sweden be a memorable way to celebrate!"

The event was held in November of 1988. During the day the wives took part in the "ladies' program" while the men attended scientific lectures. In the evening, they held a grand dinner attended by their honorary president, King Carl of Sweden. The evening affair was full of glitter and pomp. My wife and I were invited to be in the receiving line to greet the King and Queen.

For our second visit to Stockholm we had time to wander the streets. There were no young teenagers left at home to get into mischief. On our previous visit we had toured the palace and gazed in wonder at its many treasures, we had been taken to visit the summer home of the royal family, and we had been wined and dined by one of the leading Swedish families. This time we were happy to be left to explore more casually on our own.

The many anniversaries that followed did not occur against such memorable backdrops, though our fortieth anniversary was a milestone worth remembering. I was wealthier by then and able to afford the luxuries that had been out of reach to me in my youth.

I treated the whole family, our daughter-in-law included, to a grand holiday on Lizard Island, off the Great Barrier Reef in Australia. It took some planning to make everyone available for the same time slot. The rainy season began early that year, but the weather did not hamper our activities. We were snorkeling and swimming in the sea, so we were going to be wet, anyway.

It was not my first visit to the north coast of Australia. Some years earlier I had gone to Sydney in connection with university academic business. Taking the opportunity to add some days of vacation, it was convenient to stop in Cairns on the way back to Hong Kong. I had received my scuba diving certification, but had always regretted not seeing more of the reefs due to my inexperience.

Except for the winter monsoons arriving early, Lizard Island and its facilities were all as beautiful as advertised. Outside one of the more sheltered curving shorelines, the current and waves were wind driven. With a motorized rowboat, my wife and I puttered around the cove to the Blue Lagoon of Brooke Shields fame. We tried to reach one of the small islets, but the waves proved too overpowering for us to beach the boat on the rocky shore. Instead we retreated to a sandy strip with a view of the lagoon farther out. The lagoon was not blue that day, more like a muddy green-gray. The waters were still and we set up a picnic lunch and snorkeled for hours. The coral was beautiful and seemingly within reach. Yet, in reality it was still a good number of feet below me, the distance foreshortened by the effect of the seawater. So we ate our gourmet lunch of salmon and salad, complete with a checkered tablecloth and wine. The seagulls stood by in anticipation. They seemed familiar with the routine of fighting for lost morsels and crumbs.

To celebrate our anniversary and the family reunion, we asked the restaurant to serve us their special lobster menu. Unfortunately, the lobster turned out to be a frozen item and we would have fared better by selecting the regular dinner options. Apart from their special, the food was fantastic. Our week of luxury was full of happy memories.

The next big anniversary will be our fiftieth. What shape will I be in? That is hard to predict. Certainly longevity is in my genes, as my father lived to be 95 years old. Gwen's mother, the feisty old lady, died on her 100th birthday, battling her daughter to the very end. As she lay in her hospital bed, determined to end her life by refusing all food and drink, she still had the strength to pinch Gwen. By now Gwen had the courage, and fluency in Cantonese, to confront her mother, but more importantly, the maturity to better understand her problems.

"You should appreciate the care my sisters have given to you all these years. Instead you think only of your son and his attention. If only you could utter a single word of love or appreciation to your daughters; not, of course, to me, because I'm the naughty one who ran away!"

As she continued to stroke her mother's hand, Gwen suddenly gave out a loud yelp: "She pinched me! Did you see that? She pinched me hard."

It was certainly a reprimand, though no one will ever be able to fathom exactly what she wanted to say. We flew out the next day. Her mother died a week later, on her birthday, just after the nurses had brought in a cake with candles together with a congratulations card from the Queen of England for reaching her centenary.

People are living longer lives in our aging society. What fate has in store for each of us, though, is the unknown factor. Now both my wife and I are among the older generation. We are observing life from a new position. I can look forward to celebrating a fiftieth wedding anniversary, though I am not sure in what fashion—perhaps a luxury Mediterranean cruise. I will be at peace then, but will the world?

PROFESSOR KAO MEETS WITH LEADERS OF THE CHINESE GOVERNMENT DURING
A VISIT TO BEIJING AND SHANGHAI IN SEPTEMBER 1992 (*from left: Professor Mun
Kin-chok, Professor Kao, Mr. Jiang Zemin and Mr. Lu Ping*)

The Great, the Rich, and the Powerful

Being a modest man, I still find it a bit hard to believe that I belong to any of these categories: the great, the rich, or the powerful. Other people may say that I do—it is all a matter of definition and perception—and I wear the mantle with a comfortable, yet unpretentious fit that I can discard when I feel it is appropriate.

The pioneering paper I wrote in 1966 introduced the idea to the technical world that fine glass fiber threads in place of copper wires can be used in communication systems. As a consequence, the mantle of greatness fell upon me.

Greatness surely wasn't based on my ideas alone; timing was essential. If the timing for glass fibers for communication had not coincided with a particular need in industry, the idea would have died and been buried by history. Someone might have resurrected it successfully much later, but by then my contribution would have been long forgotten.

Perseverance also came into play. I believed in my ideas and hung on like a tenacious bulldog. Convinced that I must sell my ideas to as many people as I could, I traveled to Japan, Europe, and the United States. At research facilities everywhere, I learned that others were working along similar lines and were willing to hear more—though those at AT&T Bell Labs acted coolly and seemingly with little interest. Still, the more people that I could draw into my ideas, the more synergy there would be to put them into practice. That's exactly what came to pass. I was also fortunate that STL continued to fund my project and that my manager agreed that we needed to make contacts abroad. The number of professionals in the field grew exponentially each year. A group in the United States organized special conferences calling for technical papers to be given on related topics. Attending the conferences—ECOC, IOOC, OFC—I met all of those early friends from far-flung companies. We pioneers would chair meeting sessions, present our papers, hold panel discussions to answer questions, and address an audience of engineers from around the globe. The conferences, besides being informative, were also enjoyable, family-like affairs.

When I was a young engineer, I looked in awe at more senior technical staff who were entitled to add F.I.E.E, F.I.E.E.E., and other letters after their names. I was just a lowly member of these engineering institutions. How did they get elected as Fellows, a more highly prized status? I questioned my mentor as to how this was achieved. "It just comes as you get older and have more professional experience under your belt, young man. You don't just apply. If you do good work that is meaningful, then someone will nominate you. If the institution thinks you deserve it, then you become a Fellow."

I knew lots of colleagues who were older, but who hadn't received such recognition. So, what exactly did they do wrong? What was I supposed to do, other than to get old!

Still, as my mentor suggested, in time I became a Fellow. At the conferences, now grown in size and attendance, I couldn't believe that strangers wanted to greet me and shake my hand. I was the same person as before. Was I a great person now? I did not think so; though I had acquired a lot of knowledge and experience, which I applied to my work successfully. It was gratifying that my work was recognized by so many of my peers.

By now my company had patented many of my ideas and I had numerous technical papers published in different journals. Anyone who has worked for a large company will understand that the company owns all the rights to their employees' ideas, especially those that are formally patented. I would never get wealthy through royalties, as they go to the company. Wealth for me came the hard way, as the TV advertisement for a well-known U.S. brokerage firm used to announce, I . . . earned it!" Recognition did come via a nice annual salary, raises, and promotions, until I hit the ceiling of my executive pay grade. At that point, I only received lateral promotions. In that system I would have remained a backroom boy for the rest of my career. But I had ambitions of becoming a CEO one day.

Luckily, many years later, I received a phone call from a headhunter in the UK. "Your name has been given to us as the right man for a CEO job here in England. You come highly recommended. Your qualifications and experience fit perfectly for the position. It is with one of the leading companies in London."

At that time I was running a university in Hong Kong and was not interested in the least in changing my job. Yet, I was also very amused by the call, so I let the man continue talking as I sensed he wasn't sure what I was thinking.

Because I did not sound immediately interested in his offer, he became more persuasive. They would pay for me to travel to the UK and meet the board of directors; this was serious interest. I casually mentioned to him that I was Chinese. I had a sense that my friends in the UK, who had made the recommendation, hadn't passed on this information to the caller, otherwise he would never have called. There was immediate silence on the other end of the line, and then he hung up. That is what I mean by the glass ceiling.

In my lifetime I have been fortunate to meet many people of great stature. Initially this was from afar. Shirley Maclaine, who had achieved so much fame in her acting career in Hollywood, was in China at the same time as a group of academic lecturers. My family and I were part of that group. China had just opened up to the outside world after President Nixon's visit in 1972. At the May Day celebrations in Beijing in 1973, we were all part of the huge audience gathered in Tiananmen Square to watch the performances. Miss Maclaine was seated in the front row with the leading heads of the Communist leadership. I gawked and strained with everyone else to try to catch a glimpse of her.

Along with this star sighting, that first trip to China was memorable in many ways. We were among the first outsiders to catch a glimpse of Communist China, to travel on its trains across the steppes to Xi'an, where the terracotta figures still laid buried and the first visitors to fly in their antiquated fleet of planes. Like some rare species, we were given special meals, housed in the confiscated palatial homes of the past powerful rich, and allowed to tour the Forbidden Palace.

We visited many educational institutions and communes, which was the real purpose of our trip. It was the custom then for all visitors to these organizations to first hear the statistics and reports of the status quo from the leaders of their committees. These introductions would drone on for at least two hours or more before we were able to see the real work. Decades later I met a head of one of those universities we visited. By then, though frail, he still sounded a note of defiance. "I remember your group coming to the campus. I was under house arrest and none of you was aware of this as I was put on the stage to mouth Party slogans. It was farcical as none of the statistics were true! Afterwards I was locked up again."

We were treated as important people with special privileges and some of us reacted accordingly by feeling all-powerful. One professor in our group picked up a carved stonework of some antiquity. He demanded that it be loaded on the plane with his luggage, even though it was over his weight allowance. My family traveled light and had the least trouble with our few possessions.

"Well," said the professor, "then I will apply other people's unused weight allowances."

He looked at us pointedly. Gwen was indignant. We were not very popular when we refused on principle. The same professor took possession of all the official photos that were taken by our hosts and kept every one of them for himself. This was in the era when we all had to be very cautious about any pictures and exposed rolls of film had to be developed in China. Gwen wanted the photo of her shaking hands with a very senior Party official. This request was denied.

The trip opened my eyes to the pains of the developing country. Not only to the long, difficult road ahead for its citizens, but also of how far it had come from the feudal and corrupt stage it had been stuck in for so long. While the methods used to achieve the modest steps forward were abhorrent, we should remember the heart-rending struggles the industrialized countries also experienced to arrive at their present-day humanitarian values. Each of the roads passes through a differing cultural landscape and the values reached cannot escape being molded by the unique, rocky terrain of the history of each country.

I suppose it is with this unspoken sympathy for the country of our forebears that none of the group said a word about our air incident. Flying conditions and facilities for the public were at an elementary stage of development in China in the 1970s. The public did not fly; only the powerful and leading dignitaries flew. We would sit for hours in airport waiting rooms, with not a snack bar in sight, for a plane to fly in from another city. It was usually delayed by weather or by its privileged passengers. Schedules did not exist. There were no high-tech guidance systems at the smaller airports and pilots, generally ex-air force men, flew by instinct. We had already experienced the unnerving habit of the pilot coming in to land at higher speeds than on normal commercial flights, bouncing like a football on the tarmac.

On this particular flight, we had been waiting for a while in the small airport before boarding the plane. Our fellow passengers, all seated on one side of the plane in its midsection, were military personnel. The civilians, mainly comprised of our tour group, were seated in the midsection on the other side. This was the seating plan on all our domestic flights. Perhaps loading the midsection only made the weight more balanced for flight? The propeller-driven plane started up with a noisy roar. The plane bumped its way onto the airstrip, revved up, and began to speed down the runway for takeoff. Just as it was about to lift off, the pilot cut the engine and aborted take-off. The plane skidded off the tarmac, onto the muddy grass. The wing clipped the roof of a small house. A farmer, on a ladder against the house, fell to the ground with fright.

For a moment there was a deathly hush of shock before pandemonium broke out on our side of the plane. The army soldiers sat stiffly upright, stoically silent, while our side of the aisle screamed and yelled— terrified and undisciplined. The plane tilted to one side and the drop to the ground from the exit door was too high leave safely. After several hours they wheeled some steps out. We de-planed onto the muddy grass, feeling shaken and wobbly.

Back at the airport waiting room, two other European civilians refused to fly and decided to continue their journey by train. We really had no choice but to wait for another plane to arrive to take us farther. We had survived and fate was choosing to allow us to continue the trip without further mishap. When we all returned to Hong Kong and gave press interviews, none of us mentioned the incident. It was an unspoken feeling that we should not expose China's misfortunes to the world.

This was my first brush with power. In 1973, my values were still being shaped by my experiences and observations. Without being really aware of it, my exposure to power grew steadily. One of the perks of working for a large corporation was the support system. Staff made all the necessary travel arrangements for business trips, secretaries took care of communications and reminded me of my appointments and agendas, and benefits such as medical care for both my family and me were provided. They even provided a relocation service to help us settle into a new home when we were transferred out of town. As my years of service increased at ITT, my benefits grew commensurate with my seniority.

My neighbors always knew when I traveled for business. A large black stretch limousine with a smartly uniformed chauffeur would be waiting for me outside on my driveway. Sometimes the limo would take me to the general aviation section of the airport to board a private company plane. ITT had several planes that would fly regularly between New York and Brussels, where the European headquarters of ITT were located. The liquor flowed generously during the flight to make the journey easier. The seven- to eight-hour flight allowed the busy executives to be rested and ready for the morning meeting scheduled upon arrival. This was how you burned the candle at both ends. Of course, the men who could conduct business in this manner had to justify their astronomically high salary and bonuses. I worked long hours and during this time I was often away from home.

Gwen occasionally traveled with me. On one occasion they sent a stretch limo fitted with a bar and a small television. The driver apologized, as the normal vehicles were busy elsewhere so he welcomed us to help ourselves to the bar and to watch TV. She kept repeating "Wow" to herself quietly as she slid into the car. The seats were plush leather, low slung, and huge. There was even a small conference table with enough seating for a private business meeting. I would recommend such a vehicle for a long-distance journey. We sat on the two seats at the back and every time the driver braked, Gwen, in her silky clothes, found herself slipping forward off the seat on to the thick carpeting. This was before seat belts were normal fixtures in all cars.

The nicest perk was being met at the airport by a driver. As I came out of the exit doors from customs, I would scan the placards held up by the many drivers among the crowds of people there meeting friends and family. There would be one that announced in large letters "DR. CHARLES KAO." I walked towards the holder with a nod of recognition; he would hurry to take my bags, and then we would head to the car parked conveniently nearby. Without any further hassle, I could fall asleep in the car and be driven home.

I even flew on the Concorde once. That summer, I had cracked a spinal disc, the exact same bone and in the same place as my father had done in 1967—by sitting down suddenly and hard on my bottom. Actually I was really very lucky not to have had a more serious injury. Our log house at the lake was almost complete and the family was spending a weekend working on it. I was on the top of a ladder adjusting the gutter. My son held the bottom rungs to keep the ladder stable, as it sat precariously against the eaves at a dangerous angle. In a moment of carelessness, I dropped my hammer and without thinking, asked my son to retrieve it. Of course the ladder slid away as soon as he moved. I fell five feet and landed awkwardly. Fearing spinal injury, we were all alarmed. "Can you wriggle your toes? Don't move. Why didn't you roll when you fell?"

An immediate trip to the emergency department confirmed a hairline crack. I had to wear a brace to provide support to stand up and to prevent bending. Needless to say I was in extreme pain and discomfort. I could not travel to any meetings, however important, in this state. If I really had to attend a meeting in Brussels, I told them they would have to send me on the Concorde.

To my surprise, the company did just that. The flight took a little over three hours instead of the usual seven on a regular plane. The interior of the Concorde was surprisingly narrow with rows of two seats on each side of a center aisle. An indicator above the bulkhead seat indicated the Mach number so that passengers would know when we broke the sound barrier. The flight was smooth though there were slight shudders and the walls of the plane felt very warm to the touch.

I flew on the Concorde a number of times. Even when I flew on British Airways, as a frequent first-class passenger between New York and Europe, when the airline needed to bump off passengers from overbookings, I would get a phone call a few hours before my journey. "Sir, we would like to put you on the Concorde to Europe as we are overbooked on your present flight. We do apologize and hope this will not be inconvenient for you. The flight takes off an hour later than your original one. There will be no additional charge."

I was always most happy to oblige them. It saved me a lot of time and was less stressful on my body.

At one of the meetings at headquarters, I had the unique opportunity to make a presentation to the chairman and CEO, who reputedly had an impeccable memory. The chairman, in charge of this billion-dollar company, and his entourage sat around a very large oval table while the victim, namely the presenter, would stand on the opposite side facing the Boss. To my great surprise, he was attentive and friendly. His ability to cut a person to pieces did not materialize. I had a great time. He was about to retire and his successor was already sitting on his right observing the proceedings. I had it easy.

In contrast, I once joined a dinner function at the Waldorf Hotel in New York in my lounge suit when I was supposed to be properly attired for a black tie affair. My host at the table, an old friend, swore he had emphasized this point to me; I obviously hadn't heard him. I had a difficult time gaining entry to the affair, although if I had been in a tuxedo I might have been mistaken as one of the Chinese waiters.

As vice chancellor of the Chinese University, the chairman of a major Japanese electronics company honored me with an unannounced visit at my official residence. I was unprepared when his motorcade of several cars came winding up the long driveway. His bodyguards were at the ready when the chairman's limo drove up. I felt that I should have laid out a red carpet for the occasion. He came to personally thank me for the relationship that I had developed with his company. During the visit, everything moved according to a strict timetable. It was more like a state visit of the emperor. And he was in fact the person who had built the company up from scratch when Japan was in ruins immediately after the end of the WWII. Although not quite the emperor, he was an extremely well respected leader.

By the time I took up the post to lead The Chinese University of Hong Kong through its next stage of development, I was used to people at all levels of society, from royalty to corporate managers and to the workmen who came to fix my plumbing at home. I felt comfortable with everyone. Rubbing shoulders with the rich and the powerful in Hong Kong on a more daily basis proved to me that most of the individuals were genuinely nice people, but that they were very conscious of their positions.

During my first few years as vice chancellor, one of the perks was the use of the VIP lounge at the old Kai Tak Airport. The position of vice chancellor at any of the universities in the colony equated to a very senior civil servant rank, so it was necessary for us all to avoid the media. On departing from Hong Kong, my driver would take me to a separate entrance at the back of the airport terminal. Passport formalities and ticketing were taken care of by the lounge staff while I relaxed in a comfortable parlor, sipping a cup of tea and reading the newspapers. When the time came to board my flight, they escorted me to a waiting car that would drive me onto the tarmac, straight to the waiting plane. In those days it was usual for airplanes to be parked at a distance from the terminal buildings and for passengers to be bused out. On my return trip, a car would be waiting for me on the tarmac again, as passengers disembarked, to drive me back to the VIP lounge. Other passengers on my plane would look at me curiously as I rolled up in the car or as I casually walked down the steps from the plane into a car with doors held open by the driver.

I always felt as if I needed to act regally so as not to disappoint my audience. Yet I was also aware that such continuous treatment might do permanent damage to my psyche—with the power going straight to my head. Still, I knew such trappings would not be permanent; sooner or later I would return to a non-exotic lifestyle. This airport VIP perk was later modified after a revamping of the airport facilities.

My first driver was a veteran at his job. He had served the founding vice chancellor from the beginning, and also my predecessor. He knew all the odd alleys and spots he could wait for me while I attended meetings. More than just the streets, he also knew who the many leading men in the society were and their history. When I was invited out to dine, he always knew where my host lived. In short, he was a wealth of information to someone new to the community. Ah Yick was a few years younger than I, and when I first met him, his two sons were just teenagers. His wife worked on campus, too. After all these years of serving the vice chancellor, he had the good sense not to hang out with any of the other drivers, either from the CUHK pool or from the government car pools.

So he stayed around my residence when he was not driving me around, tinkering with the car, keeping it well polished. He also read newspapers or played with the dogs. He even maintained Gwen's ancient Honda. Ah Yick was an ideal staff member and we became good friends.

His best friend was the chauffeur from the other university in Hong Kong, the older, established HKU. Together, they had plans to build a house for their retirement back in their home village in China. On weekends, they often took trips there. It was on just such a short trip that disaster struck while we were away in the United States.

Ah Yick had contracted hepatitis C in his younger days and was under medical treatment, though the long-term prognosis remained uncertain. That weekend, he told his old friend he did not want to join him for dinner and carefully omitted telling him that he had been vomiting blood all afternoon. By the time his friend found him semiconscious and rushed him to the local hospital, it was already critical. The doctors operated on him to find the cause of the internal bleeding, but he never recovered, dying shortly after the surgery.

My secretary phoned me long distance and gave us the sad news.

Gwen was distraught. She tried to console herself with shopping and bought an expensive item of clothing that she never wore. Nor can she bring herself to throw it out. When we went to pay our respects to his ashes, sealed in a niche in a Buddhist memorial overlooking Shatin near campus, Gwen had to be comforted by the widow instead of the other way around.

My next driver was chosen from the CUHK pool. To be fair, I felt it should be based on seniority. Unfortunately the best qualified was a heavy smoker, so I ruled him out. Ah Hoi, the next senior, was an affable man, but not very polished; he was a man of the earth. He always meant well but he did not know how to be diplomatic or polite. I kept trying to rein in his enthusiasm and asked my secretary to speak with him, to educate him about proper behavior. He would always hang out with the car pool drivers on campus and I could never find him. I could have ordered him to remain up at the house when he was not driving me, but he was not very literate, so he would have been especially bored, and he was afraid of dogs. I could have demoted him, but I did not have the heart.

Where Ah Yick would have been discrete, politely ignoring a situation, Ah Hoi would curse the offender loudly and boastfully indicate that he had an important passenger in his car.

This man, now in his fifties, had been one of the many refugees of the 1950s who had made the dangerous swim from the southern coast of China, through the shark-infested bay, to reach the shores of Hong Kong. His mother was working in the colony already, but was unable to support him. He was a resourceful youngster and managed to find odd jobs to survive, though that meant he had little schooling. He learned to drive, which enabled Ah Hoi to find a stable occupation. He procured a driver's job at the newly formed CUHK where he stayed to raise a family in the staff quarters that were provided. His children made the next upward move in society. One son joined the police force, while another passed his accountancy qualifications and joined the staff of a leading bank. His daughter, of whom he seemed the most proud, was admitted as one of the first twenty students of the newly opened Pharmacy Department at CUHK. When I retired from my post at CUHK, Ah Hoi was a contented man.

At that time I also had on my house staff "the Major Domo," as Gwen referred to him. The university had placed him in charge of household purchases and entertainment. "You are not of the same status as you were in 1970s when you were just an academic. So you must shop at the best shops. You should be aware that what your wife wears, whom she sees, and what she does will be of great interest to the campus community," he informed us pompously.

We were appalled that such mundane activities should be the topic of gossip. Gwen took note and tried not to go down to the grocery store on campus in just any old clothes. However, Gwen is a practical person by nature, and it was often necessary to buy some last-minute item for a dinner that we were hosting. It was certainly a hassle to dress up to run to the store, and on several occasions well-dressed female academics ended up casting disapproving glances at the vice chancellor's wife.

Over the years she became used to hiring help of one sort or another, beginning with au pairs, who were like daughters, to vetting live-in maids. Yet, she thought nothing of doing the same cleaning tasks herself. Often she would be in comfortable old clothes, demonstrating to the house staff how to prepare the meal. Her helpers were always treated as equals and became friends.

The Major Domo took Gwen to the most upscale department store to buy bed linens. The prices appalled her and she didn't especially like the choices. We then came to find out that the Major Domo liked to raid the liquor cabinet during our dinner parties. He would hang around outside in the garden getting drunk and by the time the guests were leaving, he would reappear to oversee the clean up in an inebriated state. Much to Gwen's relief, we had to fire him. By that time, she was more knowledgeable about shopping etiquette in Hong Kong and was able to handle it herself.

We had two maids on staff, one was a local woman who did the cooking, and the other was hired from the Philippines. The latter came through an agency. We stipulated certain conditions, such as a high level of education. Of the two candidates the agency offered, one delivered a smiling photograph and a typed résumé while the other looked nervous in her photo and submitted a handwritten resume. Gwen chose the latter. At first Nora was extremely nervous and her hands were so shaky that she once dropped chili sauce all over a lunch guest. Luckily for us, she learned quickly and settled down enough to become part of the family. However, after two years she succumbed to the temptation of emigration. Before she left, she recommended her cousin to take her place. Yolly worked with us for nearly seven years, and we hold fond memories of her time. This intimate contact with a different culture taught me many things. What is considered sensible in one culture might not be in another, and this needs to be taken into consideration. I cannot fault the person and condemn the behavior as bad.

Yolly also had dreams of emigrating. We took her back to the States with us once to help move out of our old home. It made her realize that a life away from her roots would be a lot lonelier and more expensive. Financial gains had to be weighed against the worth of a richer and happier life among many friends and relatives. With a generous severance package, Yolly eventually chose to return to the Philippines, where she and her husband opened a grocery store and a vulcanizing business in their village. We still keep in touch with them and she and her husband are surrounded by many nephews and nieces and are very involved with church affairs.

My friends in the West are envious of the fact that we can enjoy the services of a maid even though I have now retired. This is a way of life in Hong Kong. Where it used to be only the prerogative of the wealthy to be able to afford house staff, now most middle-income families hire one also. With many parents working long hours in professional careers, someone has to be at home to take care of the household. It used to be the grandmothers that took up these duties; grandmothers now often have careers of their own that keep them busy.

Many parents in this generation did not grow up in wealthy homes, but with increased economic gains over the last decades, lifestyles, along with attitudes, have changed. Of course, for the business tycoons and chief officials of state, their entourages consist of a large group of people with several layers of responsibility, supervised and organized in an orderly way and accountable for their actions all the way up to the big boss. In this region of Asia, the big boss wields great power. The staff of his company hangs at his beck and call: "Make an appointment with Mr. Li for dinner; book a table at the best restaurant; find the project manager and tell him to see me now; book tickets for me to go to Tokyo next Monday . . . and get me a room at the Okura Hotel." A good personal assistant gets all of this done efficiently and smoothly. The boss is shielded from all the little details of life so that he can be left undisturbed to mull over the bigger problems.

This is the same culture, with small differences, everywhere in the world. As an example, in Southeast Asia, where salaries are low for unskilled workers, even the not-so-rich have a maid. So the more wealth there is, the bigger the retinue of house servants. In the West, large homes are common, but the average senior managers might only hire a nanny for his small children and a part-time cleaning lady. Still, Hong Kong is a small territory and, as some have noted, a vertical city. With the huge premium on land space, a house is a luxurious commodity affordable only by the wealthiest.

The vice chancellor's official residence, where we lived for nine years, was built in the 1960s. It was designed with the downstairs for receptions and official functions, while the upstairs was the living quarters. The staff quarters and garages were set apart as a separate wing. The kitchen rooms joined the two parts together and were equipped for catering for at least three twelve-person tables. For smaller gatherings, different removable tabletops were available to seat from eight to sixteen persons. The men from the building department probably never entered kitchens. They did not advise any changes to modernize it. The Formica-covered worktops were adequate, but all the cabinets were of vintage quality. Not having visited any private homes in Hong Kong, I assumed it was still the state-of-the-art. By the time I found out that all the modern fixtures were available and in use in Hong Kong, it was too late. All the major renovations had been made.

One guest to the house commented on our home, "You should not be required to live in a house of such low quality. CUHK is doing you an injustice!"

I tried caterers serving European cuisine, Chinese banquet, and home style cooking, but found all of the alternatives lacking in one way or another. The better caterers from the city were reluctant to come out to such a distant location and the local ones were not of a high enough standard.

One solution was to hire a professional cook to join the staff. I advertised the position and gave one applicant a trial meal to cook for just the two of us. I remember with hilarity the dish of chicken with cashews that he produced. He had purchased two pounds of the nuts and fried it all creating a dish of cashews with a bit of chicken. Needless to say, he was not hired. In any case, how much work would there be for such a position? Many days I was out for dinner, I was usually not home for lunch, and I did not eat large breakfasts.

I found the rich foods we ate at the numerous dinners we had to attend too hard on the digestive system. With two or three such dinners a week, I could become rapidly overweight. Gwen came up with a suggestion: "I bet all of the guests we might be inviting over to the house are just as tired of the rich foods they're forced to constantly eat. Why don't we try to serve good, fresh food, cooked simply? I could try to train our maid to produce suitable menus."

We eventually developed relatively simple, wholesome recipes and served them in reasonable style after a few disastrous experiments. We ate healthily, and our guests seemed to be satisfied and surprised by the tasty home cooking.

The purpose of entertaining guests was not only for politeness or etiquette. These dinners helped forge important relationships, reinforce trust, and hammer out common understandings in the relaxed atmosphere of a home. I hosted many of the top officials of Hong Kong, from visiting dignitaries, local and foreign university presidents and ambassadors, to well-known people from various walks of life, such as the famous cellist Yo-Yo Ma. I kept a visitors' book to remember the who's who from my guest lists.

We had many an occasion to meet the rich, famous, and powerful people in Hong Kong and from around the world. In colonial Hong Kong, the Governor often invited us to be guests on various occasions when politicians and heads of state, kings and queens, princes and princesses, and corporate leaders came to town. We met them all. Those gatherings allowed us to observe and speak with the dignitaries personally, though fleetingly. It provided them the opportunity to gain insight into the local Hong Kong culture. Occasionally, we met for longer chats.

I met Premier Lee Kwan Yew twice, once when he was still the premier of Singapore and the next time when he retired to become senior minister. He was very approachable on such occasions. We exchanged in-depth discussions on various political aspects of this region.

On one occasion Princess Anne came to Hong Kong and the Governor hosted a dinner for her in large gardens in the central district of the Island. Gwen was not quite prepared for the formal etiquette. When the Princess reached us in the waiting line, she reacted by asking, "How is your mother?" Taken aback, Princess Anne responded calmly and with some coolness. This was a tremendous violation of etiquette.

While many of the encounters allowed me to meet luminaries from around the world, I occasionally had the opportunity to share my opinions on key issues relevant to politics, social issues, technology impacts, or local policies. I hoped to broaden the recipients' minds, even if it was only with my personal views. My time spent in industry, academia, and especially my role as the head of a university, brought me into contact with people at the top of the power pyramid. The higher the position I reached, the greater the number of rich, famous, and powerful people I encountered.

The chairman of the university council of CUHK was a self-made top banker in Hong Kong. His honesty and no-nonsense approach, together with his vast experience and sincerity, guided me well. I regard him as a mentor and a teacher. He gave his immense contributions with selfless generosity. He introduced me to many of the necessary contacts by way of the dinners he hosted. He also annually invited the council, the trustees, and the senior academics of the university to a wonderful dinner held in the penthouse of the bank headquarters building in Central. In a more personal manner, together with the other guests, we enjoyed informal hot pot meals and dinners in his home.

The top echelon civil service personnel of the Hong Kong government are generalists. The service recruits only the brightest of the applicants in any year and gives them opportunities to work in different areas to advance to the top posts. Most of them are quick thinkers and capable administrators. In the days when Hong Kong was a British Colony, all power resided in London where the overall policies were determined. The civil service in Hong Kong had no need to gain any experience for developing policies or for handling ministerial types of duties.

I expect there were many consultations between the Governor of Hong Kong, the appointed representative in the colony from the English motherland, and the ministers in London. The civil service was only responsible for implementing the decisions and instructions that were imparted to them by the Governor. After sovereignty returned to China, and the autonomous region of Hong Kong was established, the overall policy decisions and the coherence of these crucial implementation steps now fell on the heads of the civil service. It has been a painful learning process in the years since the 1997 transition. Many of the shiny ideals lost their sparkle through unfulfilled public expectations. The recent step of establishing ministerial functions headed by private sector leaders or by civil service managers, who had shown the right quality as policy makers, is an experiment in the making. This step is definitely heading in the right direction. Let us hope that within the next five years, the significant overall policies can be in position and functioning, leading to a cohesively run Hong Kong.

During my working days I met people from many walks of life. Occasionally, I was lucky to meet famous people like the first Chinese Nobel Laureate, Professor C. N. Young. I got to know him well because of his association with CUHK. His contributions to the field of physics were spectacular and his enthusiasm for physics was contagious. He is an icon and the pride of the Chinese world. The world famous cellist, Yo-Yo Ma, whose mother's sister, Mrs. Cho Ming Li, is the wife of CUHK's founding president, accepted an honorary degree from CUHK. I recently watched a TV program about him, which made me appreciate his well-deserved fame. On the occasion of the CUHK award ceremony, he took time to socialize with the students, showing his bubbly nature in a relaxed manner. We invited him to dinner and he remembered coming to visit his uncle when he was a small boy.

I suppose I am a bit of an oddity. Since retiring, I have returned to a more normal way of life and people can often find me on buses, trains, and other forms of public transportation, particularly when I am in Hong Kong. The public transportation system is unrivalled. With the numerous transportation alternatives, I discarded the thought of owning car, along with the added hassle of having to find a parking space. In contrast to the days when I was always being driven in a private car, I appreciate how the rest of the world copes with transportation. One day I came out of a banquet and found myself the only one without a waiting car. People were surprised when I walked off to catch a tram. I treasure my current lifestyle because it allows me to stay in touch with reality.

My wealthy contacts helped me to raise funds for my charity work with the university. Whether it was for educational projects or to improve the facilities, I developed a thick skin for asking the rich for large sums of money. It took skill and timing. The most expensive lunch for three rich guests was one I hosted at home. It was a low-key affair since these people were keen to help the university to develop a new project. One of them promised to give a million dollars, while the other two knew that they had been invited for a reason, but they had no idea how much of a reason. When the sum of $1 million was mentioned, the other two had no alternative other than to follow and save face. Three million dollars proved to be a fruitful fundraising conclusion, in exchange for hosting a congenial lunch. Everyone seemed in good spirits as they left.

For every new building, the government pays for part of the construction, and then the university is required to raise tens of millions of dollars, usually by naming the building after the desired name of the contributing donor. The larger donations usually are several tens of millions of dollars. The money raised tops up a private fund, which permits the university to develop new initiatives that otherwise could not be envisaged. The target donors are first identified through a detailed analysis of the situation. For example, a rosy economic environment makes the selection process easier as a show of philanthropy benefits the donor's business reputation.

Sometimes competition with other academic institutions diminishes the chances of receiving the donation. In general, the custom in Hong Kong favors the cultivation of a group of like-minded rich people to be fans of the university, rather than using the American method of raising funds from the university alumni base. The history of universities in Hong Kong is too short to apply this method of fund raising. Fortunately, CUHK historically has received generous donations, particularly from some of the leading old Hong Kong families, who were successful entrepreneurs. Their donations have enhanced many facilities that have made significant differences to the effectiveness of teaching, the quality of student life, and an attractive overall campus environment.

Many of the world's Nobel Laureates also passed through the campus to give public lectures. As the achievements of the Laureates were wide and varied in subject matter, I learned much from them in the long conversations over dinners. Milton Friedman, who won the prize for economics, and his wife were delightful company. Another Laureate had a reputation for being fond of drink. True to his reputation, his entourage could never find him when needed, and couldn't locate him when he was late for my dinner. Did meeting those who had achieved greatness help me to become great? Did some of their gloss rub off on me? Was this a case of birds of a feather flocking together, and had I become one of the birds?

Through diligent, hard work and saving, we can all become wealthier in comparison to our previous status. Wealth is a state of mind and need not be counted in dollars. In dollars I certainly am not poor, as I have worked hard all my life. Aware of the major expenditure that my father put out to educate his sons when he had little income, I was made keenly aware of my responsibilities to my own children. Unlike my father, whom I now know had to sacrifice so much to put me through school, I was lucky that monetary awards arrived just in time to meet another large educational bill.

I marvel at how the rich in Hong Kong casually send their children off at a young age to the best schools abroad. On the one hand, good education is to be prized; it can give a child a solid base on which to build his or her life as an adult. Yet parents sacrifice the joys of being a part of everyday life with their child. On the other hand, the rich are so busy being rich that often they do not interact with their children, leaving the day-to-day care in the hands of maids. So being sent off to schools abroad is not a family hardship. Their children return home polished and assured, ready to take up the reins of the family wealth and business, or to go into jobs arranged through family connections.

A past Chief Secretary of Hong Kong moved with just such poise and confidence meeting political leaders of the West. A product of schooling abroad, she would do no less than exude the assurance of her inheritance. The average middle-class community here is generally more widely traveled than their equivalents in the Western world. With their dual language fluency and their knowledge of the cultures of both East and West, they will be the bridges that link for a better understanding of global matters.

As for myself, as an old friend at Yale University pointed out, she and I belong to the last generation that spent our childhood growing up in our own cultural homeland. We absorbed our heritage as naturally as breathing, only to be torn away from it in our youth to then absorb a foreign culture. For almost half of our lives, we have continued to cultivate our understanding of our adopted country's culture. Though China is a rapidly changing society, this generation is as at home in China as it is in the West. With the rapid ascendancy of the Internet culture, and many years of political upheavals, there always will be a new generation coming on to the scene with unique characteristics, and the youngest generations, now influenced by the Information Age, will be notable for their ease of adaptation.

POSTER FOR THE 2010 WALKATHON

The Slow Death

Although I was initially angry at the exposure of Charles' Alzheimer's by the Hong Kong media in 2008, especially the way in which it was done, the outcome was very positive. Social workers from St. James Settlement, an NGO in Hong Kong, came to see us. They had been running a pro-ь ı for Alzheimer's victims that is historically underfunded, and the facility is housed in what had been an industrial building.

"We receive no funding from the Hong Kong government at all. We've tried. Their opinion is that as there is no cure for Alzheimer's, the money is better spent on other things!"

Over several months, I routinely took Charles to their facility. Their social workers even came to visit us at home. There seemed to be no other programs around that were of any assistance, or at least none that they knew of.

However, with the blaze of media publicity that Charles was afflicted with Alzheimer's, the community suddenly was made aware of the symptoms, the need for more facilities, and the fact that even a highly intelligent man could succumb to this disease.

Actually a diagnosis of Alzheimer's had been made much earlier in 2004. Charles' family was aware of the disease, as everyone suspected that his grandfather had begun showing signs of it in his early eighties. He lived until he was ninety-five.

We used to joke about it whenever Charles forgot his car keys.

"Just senior moments!" we said.

But the thoughts were always there in the back of our minds. In 2003 our family doctor suggested an appointment with a neurologist and with a geriatric doctor. The first of these specialists gave Charles the standard

Mini Mental State Examination (MMSE). This is a verbal and written test of simple questions.

"What is the date to day? Where are you now? Here are three words. Repeat them after me. In a little while I will ask you to repeat them. Now draw me the face of the clock and show me where the hands would be for ten to eleven o'clock . . ."

There were thirty such questions and Charles scored 23 out of 30. This was an acceptable score for a 70+ senior. The doctor said that if we were worried they could arrange an MRI. The neurologist was so non-committal about everything that I began to believe that maybe nothing was really wrong, and that Charles was simply exhibiting the normal signs of aging.

A few months later, after these medical conclusions, a cousin visiting his aged stepmother dropped by to see Charles. It was his annual visit and we had seen him a number of times over the years. He was a child psychiatrist with a practice in the Los Angeles area.

"You know, I think you should be concerned. I notice a real difference in him!"

He gave us his samples of Namenda, which was then a new drug for memory problems related to Mild Cognitive Impairment (MCI). He recommended that Charles get an MRI. This, together with the advise of a close friend who was a nurse, finally resulted in a visit to a neurologist in Stamford, Connecticut.

So what are the early symptoms of Alzheimer's? It is an insidious disease that creeps up on a person without much warning. It is a medical fact of life that we all age and part of aging is a very slow atrophy of the brain, beginning in the fourth decade of life that will cause us to be forgetful in old age. We forget where we put our keys, and spend an hour or two in frustrating searches. We go into another room to pick up an item and when we get there forget momentarily what item we needed. These are normal "senior moments" that we have all experienced. It is when this becomes a daily occurrence and the forgetfulness begins to include other actions throughout the day that we need to worry. We go out to an appointment and arrive at the wrong place; we get lost and do not know how to figure out the route home. We are less alert and we have difficulty following the thread of a conversation. We ask the same questions several times and forget the answers.

Of course Charles knew how to use a cell phone—a skill he would later lose—and he would end up phoning home to tell us he was lost. I routinely checked up on him to see where he was during the day.

The MRI showed a severe atrophy of the hippocampus, the part of the brain at the base of the head where logic and long-term memories are stored.

There will be a loss of words was the doctor's first statement. It will be degenerative and it will get worse—but it may not actually be Alzheimer's. It didn't really matter to us what the official name of the disease was—the symptoms and results were the same. The only hope was in the possibility that new drugs might slow down the rate of the brain's decline. So we started on a regimen of drugs that hopefully would improve his brain functions.

Charles remained cheerful, but dropped out of most of his commitments. We still played tennis daily, but even our tennis friends could see that his condition was worsening. They made allowances for Charles and encouraged him during the games, helping him pack things up after. We would then all go to breakfast together and joke and laugh. Our tennis group had been together for two decades. Over the years we all argued and made up again and again. We tolerated each other's failings and were very at ease after the many years of swinging rackets together.

One member of our club made a habit of reading the newspaper out loud. He followed the coverage of Charles' Alzheimer's and duly reported to us all of the details. The tennis group closed ranks around Charles, announcing that they would protect him. They continued to give us car rides to and from the tennis courts.

By this time Charles had resigned from all of his various posts as the non-executive director of various companies and spent his time at home, still able to enjoy reading. We joined friends to travel and to go out for meals and other events. He kept fit with daily tennis games and home exercise. The gym visits to the club had been gradually discarded. I eventually would have to give up tennis to supervise him; it was easier to play on the same days.

His comprehension became slower. He looked the same on the outside, but he was not the same. The Charles we knew and loved was dying before our eyes. A subsequent Positron Emission Tomography (PET) scan of his brain displayed the brain in different colors according to the glucose metabolism uptake of its different parts. The plaques, indicative of Alzheimer's, were clearly there in the frontal lobes.

Another neurologist, Dr. T, suggested trying an experimental treatment. The Phase III trials for this treatment for Alzheimer's was still being investigated, though the drug itself had been FDA approved for many years for use in immune deficiency cases. The side effects and toxicity had already been researched. What was still unclear were the possible benefits for the treatment of Alzheimer's.

Dr. T said that he had another patient on the treatment and that it seemed to keep the condition more stable with a slower decline. The trials so far also supported these results. The drug was expensive and worth trying for six months, he said. Then we would re-assess the situation.

So in November 2008, Charles began the twice-monthly routine of receiving intravenous injections that took four to six hours each time. At first he co-operated well and, together with the acupuncture, we slipped into a daily routine. Charles then became progressively more agitated and occasionally obstinate and obsessed. He felt the need to be outdoors and he would not be denied by anyone; no matter how inconvenient it was for someone to accompany him. His brain tired easily and he needed his daytime naps.

By May of the following year, it was decided that it would be better to move back to the U.S. to be closer to our children. There was no cure and I could only expect a worsening of his condition. I had worked through the stages of anger, depression, and sadness. It was like mourning the death of a loved one—except that the physical body was still there. The companionship and intelligence was gone; all that remained was the endless care. The injections were likely to be more costly in the U.S., or possibly even unavailable. Our family decided to stop them once we moved to California and to let the disease run its natural course. Once in the U.S., we investigated all of the residential care facilities in the area so that we would know our options for when we could no longer cope with home care.

We found a nearby day care center for seniors. It had a mixed clientele who were suffering from various medical problems. About 25 percent of the clients were suffering from dementia. Initially Charles participated and brought home various handicrafts and drawings that he had done in the classes. He was happy to go each morning.

But the loss of the injections was beginning to make itself felt. After a few months, Charles lost the ability to distinguish edge pieces in simple jigsaw puzzles. There were other signs of further cognitive loss. He needed more help to dress and undress. It certainly appeared as if the injections had helped to maintain the status quo.

Then in October 2009, the unexpected happened—a phone call came in the middle of the night. After the initial excitement surrounding the Nobel announcement, I knew exactly how we would use the windfall amount.

I phoned the doctor in Stamford who had done the first MRI. Did he know how we could get the injections restarted as soon as possible before the Nobel Presentation in Sweden. The awe generated by this so elusive award opened doors for us and we were able to get in three injections before December. We were hopeful that it would stop any further decline, though it could not bring back what had been lost in the months of no treatment of the IvGs. The cost of the injections, as the HK doctor had predicted, was double the cost of the procedure in Hong Kong.

As with the initial media attention in 2008, with the leaked information of Charles' condition, there was another very positive outcome. With the announcement of a Nobel Laureate in their area, the local chapter of the Alzheimer's Association in California came knocking on our door. Along with this recognition came many offers of help from the local Asian community and from new friends. I found myself no longer struggling to cope by myself.

The media, however, was intent on catching us unawares and at awk-wardly compromising moments. While we certainly did want the public to learn more, and to be more aware of Alzheimer's, we did not want to be embarrassed.

Our six-week visit back to Hong Kong in the Spring of 2010 made the disease even more evident. When brain cells are diminished and the electrical impulses are blocked by "plaques and tangles," what remains is very overworked. Alzheimer's patients tire easily, not from physical but mental stress.

The public events we had to attend were simply too much on many occasions. An overtired Charles meant that he would become uncoop-erative, any comprehension of his public duties gone completely. We saw the signs and decided to take the necessary action before the situation got out of hand. Time out for a good nap before an event had to be scheduled in. It was best to limit events to only one per day. Sometimes this was not possible and Charles would become very agitated.

An auditorium in the Hong Kong Science Park is now re-named the Charles Kuen Kao Auditorium. A playground in the school of which he was one of the founding governors is called the Charles Kao Square. CUHK organized a Walkatron to raise funds for a Charles Kao Scholar-ship and for publicity for Alzheimer's awareness. There was a competition for a postal stamp and a commemorative Hong Kong stamp was issued. I used every one of these occasions to raise knowledge of Alzheimer's in the community.

In Hong Kong there are about 60,000 to 70,000 seniors suffering from Alzheimer's, and there are suitable care facilities for only about 10% of this population. The community at large is generally unaware of the symptoms or the disease.

We returned to California in a quandary, unsure as where it would be best to live. The choice had been simple before the notoriety of the No-bel. The IvG's would not have continued due to the expense, the decline would have been swift, and residential care would have occurred early on—sad but inevitable.

Now we knew of a better facility for day care in Hong Kong, where the cost of a full-time helper was much less costly than in California. More friends who knew Charles well before this calamity hit kept coming to his aid. On the other hand, the California weather is very comfortable and we were slowly getting to know the community. Help was also available, though at a greater cost.

I then discovered an Asian day care for dementia patients in San Jose. Charles was no longer participating as well in the activities at his original facility. The social worker reported that he preferred to walk away and observe what was happening around him, rather than join in the activities. The Asian day care was a long bus ride away and it took several weeks before I managed to get Charles enrolled into the outreach program to provide the rides that he needed.

In the meantime, I got the bus with him and rode into San Jose three times a week. The bus trip took one hour each way. While he attended the morning sessions, I spent my time in the wonderful facilities of the Martin Luther King Library. Some retired professors from Taiwan arranged to pick him up twice a week for some socializing and suitable activities that Charles could enjoy. This meant two afternoons of relief for me.

With one to two mornings of day care, twice per month mornings taken up with the IvG transfusions, and Friday afternoons of lunch and ping pong at the local senior center, most days of the week had events for Charles to be out of the house, either with me or other caregivers. Even with all these programs, the remaining hours of the day were stressful, as constant supervision was becoming a necessity.

Charles proved he was still able to travel and to cope if his minders knew his needs. He was invited to celebratory conferences in both London and Paris in the summer of 2010. It was the 50th anniversary of the invention of the laser. Without the development of a suitable laser to use in conjunction with the optical fibers, a transmission system would not have evolved.

I had been reluctant to accept the invitations; the organizers seemed unconcerned that Charles would be unable to give a talk. They wanted him to be present regardless. A relative was able to accompany him and the organizers were willing to pay for this, so the trip was on. I applied the various coping tactics with Charles that I had picked up from our spring visit to Hong Kong. Vague memories of the many long business trips to far-flung locales were lodged somewhere in his brain. Charles had no trouble sleeping in the string of strange hotel rooms. They seemed like second homes to him and he found the bathrooms in the middle of the night with little problem.

He performed admirably at all of the functions; the ceremony for an honorary degree from University College in London, the one-day lectures held at the Royal Academy of Engineers, and a small dinner held in the House of Lords. During this London visit, we also learned that Charles had been named in the Queen's Birthday honors list to be knighted!

Amazingly, his childish scrawl in an autograph book, announcing his hope to become "Sir Charles," had finally come true.

The need for sufficient rest for Charles during the day and the fact that he could not be left unattended meant that all of the arrangements revolved around his naps. We did very little sightseeing. In London, we walked in the nearby parks for exercise, and we did manage to make short visits to the British Museum and the Science Museum. We arrived in London just in time to enjoy watching part of the Trooping of the Colors on the Queen's birthday from the privileged podium of the Royal Society which overlooked the Mall. The flyover at 1 pm came directly overhead, led by a Spitfire.

The Paris events went smoothly, too. A great deal of time was spent finding bakeries and supermarkets. As the accommodation provided was in a serviced apartment with a small kitchenette, we cooked some meals. It was a rather down-market accommodation, and after complaining, we were moved to the Sofitel Hotel for the last two nights—from rags to riches. All Charles saw of Paris was Notre Dame and a walk along the famous Champs Élysées! The Louvre was so crowded with tourists that after viewing the Venus de Milo in a scrum, Charles decided that he wanted out.

Most of the other events that come with being a Nobel Laureate had to be turned down. Charles was simply unable to contribute to the discussions or to give lectures. It is sad that this honor came too late in life for him to enjoy the glory.

Since the invitation to come to Hong Kong in the spring, a team had been working continuously on setting up a foundation to raise funds for Alzheimer's in Hong Kong. Alan, one of the foundation's director's, carried the team with his enthusiasm, and a great deal had been accomplished, even with me back in California. A major PR event was scheduled to coincide with the release of the commemorative postal stamps in September 2010, as September 21st is International Alzheimer's Day, celebrated by 70 countries around the world.

We were needed at a launch of the Foundation for Alzheimer's Disease, so we decided to return to Hong Kong again, with much new knowledge gained. CUHK enticed us with the offer of an apartment on campus and friends and alumni encouraged us to return, too.

My son's suggestion is to give it one more try and split our time between the U.S. and Hong Kong, until Charles is unable to travel or to cope with the long flights. Then we will need to decide which location will best suit us. In the meantime, I continue becoming more involved in the advocacy for spreading knowledge about this terrible disease, with the understanding that the road ahead will be an increasingly difficult journey.

MR. AND MRS. KAO WITH THE NOBEL PRIZE

The Most Prestigious Prize in the World

"You've won the Nobel Prize for Physics!"

When the phone rings in the middle of the night it is usually either a crank call or an emergency. Though we had known for years that the Nobel announcements were released to the media every October, we had never expected a call from Stockholm. Over the last several decades, friends had been constantly advocating for Charles to receive this honor, and he constantly reminded them that the Nobel had no category for research work in applied physics. His work on optical fiber was based on engineering concepts and basic physics. He never dreamt that his research would lead to such a tremendous change in lifestyle, only that it would alter the world of telephony and increase the amount of data that would now be able to flow through those fibers.

Charles was groggy and not in a particularly receptive state of mind. "Hunh, me! Yup, very prestigious!"

Charles promptly fell asleep again, as he had done more than two decades ago when the phone call came in the middle of the night informing him that he had won the Marconi prize. With the time difference, the Nobel announcement made the 6 o'clock evening news in Hong Kong and it was early morning breakfast conversation on the east coast of the United States. By the time we got out of bed, our phone in California was ringing nonstop. The local Chinese media was in a state of hysteria. Could they come over to interview us right away! Would we give our immediate reactions over the phone!

Ironically it was the swift transmission of data along these ubiquitous glass fibers that was enabling the instantaneous and simultaneous diffusion of the worldwide announcement. Back in Hong Kong, the public immediately claimed ownership of Charles. They were celebrating in the streets and at numerous dinner parties. CUHK held a rally the next morning on campus for their students to celebrate and to sign a huge congratulatory placard. We were glad we were not in Hong Kong when the announcement was made or we would have been mobbed. Anyone who could get in front of a microphone and camera pronounced their close-knit association with Charles. All sorts of facts, both true and false, were published in the newspapers and repeated later each night on TV. The Chinese TV crews based in San Francisco and Los Angeles filmed and interviewed us countless times. Our faces became familiar to the 7 million-plus citizens of Hong Kong and beyond, and intimate details of our lives were broadcast, including the almost continuous reporting about the details of Charles's Alzheimer's.

The first few weeks of the award were a blur of film crews. Packs of reporters were hanging around everywhere. Recording vans were parked along the road to our house. At least our neighbors seemed impressed, especially the evening that the blazing spotlights for one of the interviews overheated our home, which set off the fire alarm. The shrill piercing sound brought out all of our neighbors. I frantically dashed around the side of the house to turn off the alarm at the control box, which brought only momentary peace.

Five minutes later the alarm went off again. The local fire station was just at the end of our road and when the alarm sounded for the second time they were at our home within minutes. Charging in with neighbors in tow, and no fire in sight, they quickly made sense of the situation. Thankfully they were amused by the false alarm, and gave their congratulations as they headed back to the fire station.

TVB, a Hong Kong station, wanted to come back to interview us again and again. They had a permanent crew based locally in California. It would be nice, they suggested, to make a longer documentary about Charles's life. I was growing progressively angrier at the intrusion. Our son berated us for being so naïve.

"Why are you guys doing this all for nothing. They're making money off you. You should have an agreement and a signed contract!"

I thought about this and then wrote a letter asking for donations to the local chapter of the Alzheimer's Association in the U.S. and for an equal sum to be donated to an NGO in Hong Kong, if we were to grant any more interviews. I addressed the letter to Ms. Fong, the wife of the top executive at TVB. We had met her and her husband on numerous occasions, and I felt relatively comfortable that she would respond positively.

A reply came thanking us for the filming that had already been done. TVB now felt that they had enough footage and would not need to trouble us in Sweden during the ceremony. In other words, there would be no donations to a worthwhile cause.

My friends all responded, "Didn't you know that Ms. Fong has a reputation for being very tight-fisted. You will never get a penny out of her!"

With the constant stream of Chinese TV stations interviewing and filming us, different snippets of footage were shown every night throughout October, to the extent that even the Chinese community here in California recognized us in the mall. People wanted to snap pictures of us on their cell phones—yet another gadget made possible by optical fiber.

The first indication that the U.S. media had any interest came with a phone call from the Mountain View City Hall. They wanted to honor Charles with a plaque and to name November 4 Charles Kao Day! The City Hall of Santa Clara followed with another plaque, as did the Senate in Sacramento. All in all there were about five or six such plaques, which we decided to display in the garage of our town house in California. Our home is small so there's not a lot of free wall space. Now every time we go in and out of our home via the garage, we can admire the plaques.

In general, the major American media outlets seemed unaware of where to find us and showed no interest beyond the press release given to them by Sweden. That was a blessing. Eventually a local crew from CBS called and came over to film. The interviewer told us that his mother also had Alzheimer's, so he understood. His clip was broadcast just before an advertisement for the Alzheimer's Association.

News of Charles's Alzheimer's was broadcast far and wide early on, so the Nobel Foundation was well aware that this particular Nobel Laureate would not be able to give the required speech in Sweden. I was the obvious stand-in, but I had never spoken to a large audience before. Could I do it? Could I even write the speech? The Nobel committee informed us that in a previous year they had had a similar situation. Our son refused the task, as did several professors at CUHK, all of whom insisted that I was the proper person to deliver the speech, and that if I decided I wanted assistance putting it together, staff from CUHK would be there to help.

I had edited all of Charles's speeches over the years. His English, though good, was never as fluent, because I had had rigorous English grammar lessons at school in London. "Its the prepositions," I wrote in the preface of one of Charles's books. "He cannot use them correctly!"

The media continued to hound us, and we eventually learned not to answer the phone and to allow the answering machine to pick up. Whenever the recording was dead silence, we knew it was the media.

I spent the next several weeks working on the speech. The technical parts of the speech were the thorny issue. Charles had forgotten almost all of his technical knowledge, which made the assistance of various CUHK professors necessary. They wrote and re-edited parts of the speech, putting in many hours of work long distance, with e-mails flying back and forth between Mountain View and Hong Kong. Between the two venues, the slides for the talk were chosen and it all came together at several meetings in Hong Kong.

During my absence, our son moved back home to take care of Charles. We were also lucky that a friend had arranged for retired professors to meet with Charles on a daily basis. This enabled our son Simon to work from home and then to run into his office for meetings. Day care was also an ongoing activity, but that only covered the mornings. So the "professor" sessions were helpful and very much enjoyed by Charles.

The beginning of the celebrations began with a trip to the White House in November. As I was still in Hong Kong, I missed meeting the president. Charles and Simon traveled to Washington to meet President Obama, together with the other American Laureates of 2009. Simon remarked that he was glad it was not the previous president; he had not voted for Bush. That same evening there was a formal dinner at the Swedish embassy including guests from the relevant government departments and several senators.

With the numerous congratulations came just as many requests from friends and relatives.

"We would love to join you in Sweden. Count me in!" Before I had a chance to really take it all in, we had agreed to invite a large number of people before finding out that we were only allowed to bring up to fourteen guests to the events. We hurriedly excluded spouses from the list and fit in a few close friends. We were then informed that if spouses still decided to come to Stockholm, they would be unable to see any of the events, except for news coverage on TV. Sweden is quite expensive and the officials suggested that the extra guests would be disappointed. As for the long list of ex-colleagues who had worked with Charles—whom amongst them could we single out to invite. This was a daunting exercise that we avoided by restricting the list to close relatives and friends.

To kick off the celebration, a Sunday reception of just the Laureates and their family members was held in the Nobel Museum. We listened to a short history and were given a tour of the exhibits before we sat down for lunch. Galileo's original telescope is housed in the museum. Also of interest is the overhead continuously revolving line of portraits of past Nobel Laureates. We were told it would take two hours to view all the portraits. And each chair in the canteen has the signatures of past Laureates underneath.

The curator asked me for any of Charles' original notes made while he was working on the optical fiber research. We have moved so many times and thrown out so much paperwork each time that there was very little that we had kept. We never dreamt that we would be part of history. Someone suggested that there must have been lab notes kept from the time, but none were found. Richard Epworth, a former colleague, had been sufficiently foresighted to keep all of the soon-to-be-memorabilia. Amongst his stash was an ancient video showing Charles working at his bench and a sample of the circular wave-guide that might have been used for telephony instead of the optical fibers. Richard assumed that the lab notes had all been turned over to Corning during the deposition, when controversy had arisen as to who owned which patents.

The official portraits of each Laureate were taken later the same day. For dinner, we went to a restaurant in the Old City, Gamla Stan, for some traditional Swedish cuisine. The black limo was parked in a nearby side street and when we retrieved it we found someone had unpeeled the plastic Nobel logos on each side of the car doors.

As Alzheimer's had now limited Charles's vocabulary, it was obvious that the press interviews were not going to happen, and we were able to excuse ourselves from them. This is why Charles is missing from the group photos of the Laureates. A full day of activities began on Monday, December 7, 2009, with a breakfast meeting at 8 a.m. at the Royal Swedish Academy of Sciences, followed by a press conference. We joined in after these events for the rehearsal at Stockholm University. In the large Aula Magna hall, the Laureates ran through their Nobel speeches, including me.

I had sat through so many talks given by Charles, and after years in the public eye as the wife of a university president, I had gradually grown used to the exposure. Nevertheless, I surprised myself by not feeling nervous. There had not been much time in all the rush to practice, let alone memorize the talk. After the brief rehearsals, we had time to return to the hotel and rest before an evening reception and dinner at the Royal Academy of Sciences.

The Royal Academy of Sciences building is notable for all the portraits of the many great scientists hanging on the walls of the wood paneled rooms. There was a group of young students from Asia attending with their teacher. They had been picked to attend as a reward for scholastic achievement—bright, sparkling kids, somewhat in awe of their surroundings.

The next day, some of our guests arrived in Stockholm in time to hear my lecture. I felt fairly composed, no nerves, and spoke in a measured voice. I kept my eyes down most of the time, as I was still unfamiliar with the text. Knowing that the Chinese University was broadcasting the speech live in Hong Kong—where staff, students, and the public were gathered on campus watching, I decided to send out a greeting to them.

"Distinguished Guests, Ladies and Gentlemen, Relatives and Friends,

Good Morning," I began. Then I added impromptu in Cantonese, "Greetings to you all!"

We heard later that this was greeted with a loud appreciative cheer by the audience in Hong Kong.

I proceeded to give the rest of the talk smoothly, eliciting a few laughs along the way. Charles sat in the front row, nodded with approval, and smiled. The other two Nobel Physicists gave their speeches and then it was photo op time, along with a general milieu of congratulations from everyone. We were free to leave after lunch to relax and get ready for a reception before the Concert in the Stockholm Concert Hall. After all this excitement we were more than ready to take a nap. Our two children, younger and more full of energy, went off to explore the city with relatives.

Lena was our limo driver. She drove us to all the venues, waited for us each morning as we were ushered out into the grey day (there was one day of sunshine and no snow in Stockholm), and opened the doors with a smile as we slid into the low comfortable leather seats. The car was spacious enough to hold all five of us when necessary—the two of us, Simon and his girlfriend, and our daughter. At the end of each event, there she was, just as cheerful and as welcoming. The organizers also provided us with a minder. She ensured that we arrived on time, found the correct entrance doors, and shepherded us to the correct sites. Eva was as tall and blonde as Scandinavians have a reputation for being. Without her by our sides to whisper protocol and other words of advice, we would have been totally lost and bewildered.

At the concert, the lovely Princess Victoria and her party were seated just a few seats away from us. She turned and smiled several times. She is the very model of a Princess—charming and gracious. After the concert Charles and I attended a dinner at 10 p.m. held in the Grand Hotel. The rest of our guests went off and had a big party themselves somewhere else.

Not surprisingly, we slept in late the next morning. We were free until a lunch hosted by the British Ambassador at his Embassy. As we hold dual citizenship from both the U.S. and UK, we had a choice. There were several Nobel Laureates from America, so we chose to be entertained by the British, where we would be the main guests of honor.

We heard from the ambassador that the British Embassy in the long past used to be a rather shabby residence and the King of Sweden at that time had been invited to it for dinner. As a result, the King thought a grander place was in order and, ever since, the Embassy has been situated in an elegant house. I do not remember what we ate, only that something on the menu intrigued Simon. He e-mailed the ambassador later to find out if he could have the recipe, but he never received a reply. There must have been some secret ingredients.

That Wednesday evening a reception was held in the Nordic Museum. I wanted to spend the time looking at all of the interesting exhibits. The building itself was a marvel of gothic arches and stone walls. Instead we were surrounded by throngs of distinguished guests, local school children, and students from various countries who had been invited as a reward for scholastic achievement. Congratulations showered down on us from all quarters and we beamed back at everyone.

After the reception we were invited to dinner by the Taiwan Minister of Science. He invited everyone in our party—including all of our relatives and friends—to a Chinese banquet in a restaurant in a distant suburb. We went in our black limo and invited our driver, Lena, and our minder, Eva, to join us. We all had an excellent time and the food was quite good.

They next day was the big day—the Nobel Prize Award Ceremony in the afternoon officiated by the King of Sweden to be followed by the formal white tie dinner in the Grand Blue Hall. The formal Nobel dinner is an important engagement for the citizens of Sweden. The menu is planned months ahead and kept a secret, as are the entertainment and all of the floral themes. Leading citizens from many walks of life receive an invitation to participate. The tickets are expensive, but probably few would turn down the invitation to attend such an annual event, as the Hall seats only 1,400 guests or so.

A rehearsal was held in the morning in the Concert Hall, which was the venue for the presentation ceremony. The flowers decorating the stage were exceptionally beautiful, but if we thought they were splendid, we would be bowled over by the ones we would see later that evening in the hall. According to tradition the Physics Laureate leads in the procession. We were unsure Charles would walk in the right direction, and a professor from the university was selected to walk by his side to guide him. We were also unsure if Charles would remember the procedure for accepting the award. The King was warned of this problem. Then we were let loose again to relax until we dressed up in white ties for the afternoon event. So off we went to join the relatives for a bite of lunch together.

After a short rest we nervously started dressing for the presentation. The rented suit fit Charles well. This had been mail-ordered from the Swedish agency with the required measurements a long time ago, including the patent leather shoes. One of our guests who lived in Hong Kong found it less expensive to have the suit made in Hong Kong than to rent it in Sweden. Simon succumbed to the idea—except that he had to rely on sending his measurements from California. We figured that they both could use the suits for another important occasion—a wedding! Another Hong Kong guest did the same. He said he planned to wear it to his daughter's wedding, even though she was only nine years old. After everyone was dressed, we took photos to memorialize ourselves in our finery.

As we descended into the hotel lobby, we found ourselves the center of attention. There were many others all dressed up for the occasion, a sparkling sight of glitter and dignity. Eva was hovering, as one of our party rushed off for a forgotten item, anxiously worried we would be late.

Outside was a convoy of long black limos and a crowd of fans wanting autographs. We had not rehearsed Charles in time to cope with such autograph hunters and he either dug in his heels and refused to sign or took a long time struggling to write one signature after much prompting. He felt the embarrassment of his inability to write. It was much easier to deal with the piles of fan mail that arrived at the Hotel as Charles could take his time to sign. He got better if he could practice. Learning from this experience, our contact in Hong Kong suggested that we have a replica signature stamp made and that she would ensure it was stored safely.

It was then on to the concert hall. Again, by tradition our black limo led the procession. Police motorcycles escorted us. The roads were closed and isolated bunches of people waved as we swept by. The seating for the ceremony sent Simon's girlfriend, not yet a relative, up to the balcony seating with our other friends. Relatives were seated in the front row downstairs. Somehow Charles's brother and his wife ended up at the end of the row after his uncle and first cousins.

Simon, being unhappy with the separation from his partner, rushed up to the balcony to exchange his seat down in the front row for a balcony seat of one of our friends, Alan. Time was at a premium as Alan ran down, only to find Simon had been rejected by his girlfriend and told his proper place was to be seated down with his mother, so Alan barely got back upstairs again before the fanfare for the Royal arrivals sounded.

As the trumpets sounded we all stood as the Swedish Royal family entered onto the stage. Princess Victoria, heir to the throne, has a younger sister and brother. They were seated behind their parents and the Crown Princess. It is usual for the Royals to enter last, but for the Nobel Ceremony it is the Laureates who enter last, filing in the left side of the stage, led by Charles and his professor escort, to a standing ovation by all present.

Next began the formal presentation, with an opening speech in Swedish, which, of course, was quite incomprehensible to us, and then came the first citations. As they read out the physicists, Charles stood up after a nudge from Willard Boyle who was sitting next to him. Slowly, as if a little uncertain, Charles began to walk to the center of the stage where a big circle containing the letter N was patterned into the floor covering. The King, who was walking much faster, was upon him before he had taken more than two or three steps. Smiling, he stretched out a hand. Beaming broadly, Charles shook the proffered hand and accepted the gold medal and certificate held in a red binder. He bowed slightly to the King but forgot to turn to bow to the foundation members on the one side and to bow to the past Laureates standing on the other side of the stage. He turned a bit and half nodded to the audience, then sought the safety of his chair.

It was an impressive ceremony. The trumpets blew for each Laureate as they stepped forwards to walk to the middle of the stage and as they shook hands and received the medal and certificate. The Hall was filled with an audience dressed formally in white ties or long gowns. The banks of flowers set before the stage had been specially procured from far and wide.

There is only a short break between the presentation ceremony and the banquet. At the end and after the Royals had left, everyone came onto the stage to take photos and the crowd milled around us. Eva took care of the medal and certificate as we shook numerous hands and posed happily for many pictures. Then we were all whisked off—we in limos, and our guests on buses—to the city hall.

An inner courtyard beneath an archway leads to the city hall. Having disgorged everyone outside of this, we walked over cobblestones, past school children lining each side of the walkway, who were waving burning torches. We smiled and waved back to them. They smiled back shyly, and a few of the braver ones waved.

Inside the building, a small room is set aside for the Laureates to spruce up, remove their coats, and recover for the next exciting part of the program. There were many large entrance halls in the building, and we were gathered in a specific order in a long line to make our grand entrance into the banquet hall. The Royal family was each partnered with one of the Laureates. Due to our Alzheimer's issue, we two stayed partnered. This must have been the grandest entrance we will ever make in our lifetimes.

By now all the guests had been seated in the banquet hall awaiting the entrance of the VIPs. The hall has a very high ceiling—at least three stories tall. A narrow open gallery runs along the wall on one side and a corridor hidden from view runs along the remaining three walls. At one end of the hall, the open gallery has a grand wide stone staircase that winds its way down to the main floor. This is the stairway we processed towards from one of the outer halls. With trumpets blaring, we slowly stepped with dignity towards the audience standing at their tables, led by the King and his partner Laureate.

Down the middle of the enormous hall was a long, long table. The gracious table settings are traditional and a replica of the setting can be viewed in the Nobel Museum. Each couple from the procession went alternately to the left and the right to their allocated seats, with the Royals being seated in the middle. The other diners were at tables fanning out on either sides of the long central table. Our relatives and guests were at a table near the bottom of the stairs and had a close view of us as we descended. Although no photos were officially allowed, many of them had cameras and snuck photos of the proceedings.

Each column along the sidewalls was decorated with huge garlands. The flowers set upon the table also were wonderful. In fact, as we had savored the atmosphere while waiting for the food, one had the distinct feeling of being in the Garden of Eden. Everywhere we turned, our eyes feasted on bouquets and garlands of grand beauty and artistic arrangement.

The ceremony of serving the food was another amazement. With more trumpet fanfare, two lines of servers, marching in formation down the stairs, trays of food held high on the palms of their hands, came down into the hall, serving first the main central table and then the side tables. It was all done efficiently and swiftly. We were told that the kitchens were down in the basement below and that the trick was to get the food up as fast as possible so that it was served hot.

About halfway through the meal the entertainment started up. It was a creatively sung modification of a type of Romeo-and-Juliet tale. The villain in the form of a dragon, who had to be vanquished, appeared briefly, peering through a window that opened above from the corridor wall. The dragon would suddenly show up later on the landing of the staircase only to be chased off. This landing space was the stage for the theatrical troupe as was the staircase. To begin the play, troubadours entered from various doorways behind the garland-draped columns and strolled amongst the tables with medieval lutes. A chorus of ladies in medieval dresses sang their messages to and fro, teasing the troubadours. The actions took place from all sides and we did not know which direction to watch. The actors all gradually converged onto the stage and stately dances, songs, and music entertained us as we watched. We were entranced by the magnificent silken costumes of varying hues of orange that shimmered in the play of lights in the darkened hall. It was a feast for the senses. We were transported into a fantasy world and it was over all too soon, as the lights came back on and a line of servers appeared with the final courses.

The diners seated at the long table now arose and we exited the hall in the manner in which we had entered. As everyone stood, we formed our lines and as the trumpets again sounded, we walked with great dignity and laughter, up the grand stairway. We were led along the gallery to another long anteroom where we were told we would have private group audiences with the King and his family.

In the meantime, the huge number of diners was gradually dispersing. We learned later from our guests that there had been dancing in another vast hall—the Golden Room. They wondered where we had gone as they looked for us amongst the crowds. We had been waiting our turn to mingle with the Royals. Each Laureate got to meet and speak with the family for a few moments, and then have a photo taken of the event. The King had no recollection of having met us thirty years earlier at the Ericsson Awards held in Stockholm, which are patterned after the Nobel programs.

After this we were free to look for our guests. Most of them had already left and we only found our children. We managed to dance a little in the Golden Room and mingle with the actors in their bright orange gowns. Charles held up very well and seemed to be enjoying himself. We found our way to the student entertainment and learned that they had spent the entire year concocting a theme for their event—the human body. One area depicted bones, another had a slew of streamers arranged across the ceiling for the lungs, and so on. The room at this point had become more of a beer-drinking fest where only a few people were dancing. We took advantage of the empty dance floor and took over the space with verve. Another Laureate and his family also had found their way to the same venue. We joined forces and traded partners.

Just after midnight we were drooping, so we sought our way back to the Grand Hotel in our limo. The festivities would go on all night long, but we were too tired to muster any more energy. Although the week was winding down, there was still the King's banquet to come the next day, held at the Royal Palace. Before this, all of the Laureates and their guests were treated to a visit to the Vasa Museum and a bus tour of Stockholm.

The Vasa was designed and built in the shipyards in Stockholm during the Thirty Years' War of 1618–1648. It was the largest wooden ship ever built with an extra cannon deck ordered by the King. It was launched on its maiden voyage in the waters outside the Royal Palace. A wind caught the sails and the top-heavy vessel keeled over to port. Water rushed in and it sank in less than twenty minutes into its launch. Many unsuccessful attempts to rescue it were made over the next fifty years. It was forgotten about until a successful recovery was made in 1961, 333 years after its futile maiden voyage. After numerous years of restoration the ship is housed in all its seventeenth-century glory in the Vasa Museum, a tourist attraction well worth a visit.

On returning to our hotel, we all went to eat a quick lunch at the nearby Nobel Museum, where we found many of our relatives had also had the same idea. The food in the café is both excellent and inexpensive. In the afternoon we were scheduled to go to the Nobel Foundation to sign off for the medal, certificate, and a requested replica of the gold medal. In comparing the replica to the real medal, the latter is more golden and heavier. Without having the two side by side to compare, it is easy to mistake the replica for the real thing.

Then in the evening would be the banquet hosted by the King and Queen of Sweden in their palace. They had an apartment on the top floors of the immense building. The rest of the palace included staterooms that were used to host events for visiting dignitaries. The Royal family usually preferred to live most of their time in the smaller summer palace in Drottningen. This banquet was held in a long room and there were relatively few guests. Apart from the Laureates and their spouses, the other guests were diplomats and various government officials. Hence they all were on nodding terms with each other, holding animated conversations in Swedish. The servers standing behind us wore intricate period costumes. What we saw of the palace confirmed that it was not used much. Charles needed to use the restroom and we were escorted through the kitchens to a service toilet. The kitchen staff stared at us in surprise. No one is ever supposed to leave the table during dinner with the King, though someone else in previous years had broken this etiquette, so we were not the first.

Many years ago—perhaps thirty—we had visited the palace to see the Royal collections. These rooms are open to the public. Did we have dinner there for the Ericsson Prize in 1976? My memory of a royal dinner from then is in a smaller square room, seated at tables also set in the form of a square. It might not have been in the palace.

We flew from Stockholm early that Saturday morning, after an exciting week of Nobel celebrations as the first flakes of snow were beginning to fall. The depths of winter were about to begin.